Biomes and Habitats

BIOMES AND HABITATS

DR PHILIP WHITFIELD
DR PETER D. MOORE
PROFESSOR BARRY COX

MACMILLAN REFERENCE USA

GALE GROUP

THOMSON LEARNING

New York • Detroit • San Diego • San Francisco
Boston • New Haven, Conn. • Waterville, Maine
London • Munich

A MARSHALL EDITION

This book was conceived, edited, and designed by
Marshall Editions,
Just House.
74 Shepherd's Bush Green,
London,
W12 8QE

Copyright © 2002 Marshall Editions Ltd.

First published in the USA in 2002 by Macmillan
Reference USA, an imprint of the Gale Group

Macmillan Reference USA
300 Park Avenue South
New York, NY 10010

Editor	**Jinny Johnson**
Text Editors	**Anne Kilborn**
	Gwen Rigby
Assistant Editor	**Jazz Wilson**
Managing Editor	**Ruth Binney**
Art Director	**John Bigg**
Art Editor	**Lynn Bowers**
Picture Editor	**Zilda Tandy**
Production	**Barry Baker**
	Janice Storr

Library of Congress Card Number: 2001094842

ISBN 0-02-865633-4

10 9 8 7 6 5 4 3 2 1

Originated in Singapore by MasterImage
Printed and bound in Italy by De Agostini

The Publishers would like to thank the World
Conservation Monitoring Centre at Cambridge
and Kew for their invaluable assistance in the
preparation of the sections on Threatened plants
and Endangered species (pp 200-13) and Pip
Morgan for compiling the information.

Contents

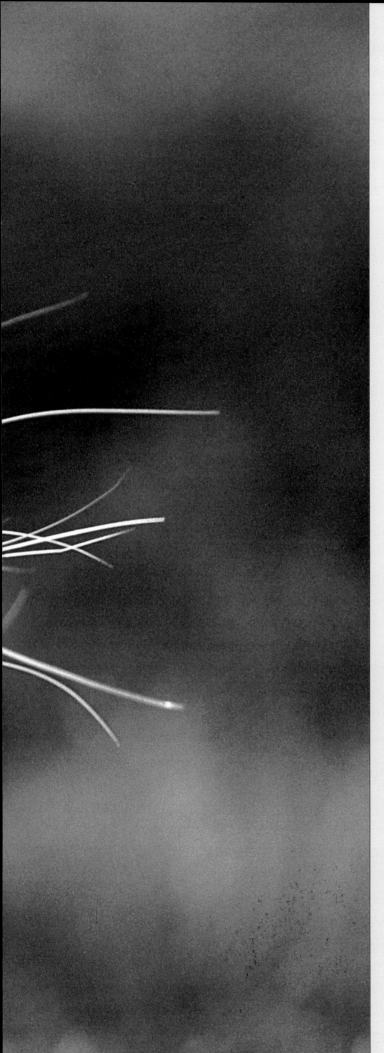

Foreword

The ever-changing patterns of life on earth are revealed in the pages of this book. It explains where plants and animals live—and why they exist where they do—and traces the vital geographical dimensions of natural history in maps. On this cartographic foundation we have drawn together the strands of modern knowledge to create a comprehensive study of the natural realm.

On a global scale, we chart the physical forces that have shaped the earth and the biological processes that have determined the life of our planet.

Coming in closer, the environment is brought into sharp focus through its many varied habitats. From tropical and temperate forest, through savanna and scrub, ocean, shore and island, desert and tundra—each has its own unique identity created by the living things that flourish within its boundaries. Comprising each habitat are the niches where creatures are engaged in the daily round of living, where resources must be shared, mates found, and young reared.

But the living planet is not static. Superimposed upon it are changing patterns of existence—of migration and colonization, of natural population explosions and crashes—which give the biological world its own dynamic. The arrival of humans has also effected great changes in the living environment, many of them documented , explained, and evaluated here.

This book reaches out to the leading edge of scientific knowledge, to unravel the vital forces that are molding the future. In doing so, it addresses the most crucial questions of planet management in to the next century.

THE EDITORS

Global patterns

The shape of the living world

It is often said that the sheer variety of life on Earth is astounding. So it is. But when you come to think about it, the *similarity* between animal groups worldwide is also extraordinary. In Europe there are frogs and lizards, butterflies and spiders, earthworms and snails. Travel to the island continent of Australia, ten thousand miles away on the other side of the globe, and you will find there animals that, without any question, also belong to such groups. It is true that Australian frogs are not exactly the same as European ones, but they are so similar in the fundamental details of their anatomy that you cannot doubt that European and Australian frogs are related and that furthermore they belong to the same group as frogs in Africa, Asia, and South America. The same can be said about the lizards and earthworms, butterflies—and indeed, most major categories of animals alive today.

A dramatic theory proved

Relationship implies that all the animals concerned are derived from common ancestors. But if that is the case, how and when did the descendants get separated from one another? Many different answers have been proposed to that question. Birds, of course, are hardly a problem. They could have flown from continent to continent. Perhaps, then, they were responsible for carrying seeds and eggs and even tiny animals in the mud on their feet. Perhaps flying insects and spiderlings also traveled by air, being inadvertently swept up by gales and blown away over vast distances at high altitudes. But those explanations cannot account for the distribution of bigger animals such as lizards. So perhaps the level of the ocean was once much lower than it is today and animals were able to walk or wriggle from one continent to another. Maybe volcanic eruptions built chains of temporary islands linking continents and some animals that were normally landbound were able to swim sufficiently far to get from one to another. Such things may have happened in particular cases, but neither separately nor together can such events provide an adequate explanation for the astonishing ubiquity of most animal groups throughout the world. There has to be a more universal explanation. In the 1960s it was discovered.

In that decade a suspicion that had been growing in the minds of many geologists for a long time was finally proved. At one far distant time in the Earth's history, all the continent were joined in one great supercontinent. About 200 million years ago, that vast landmass began to break up. The fragments, today's continents, then drifted apart, each carrying its own complement of animals and plants. The proof of this dramatic theory came primarily from geophysics and other branches of geology, but its implications have now affected almost all of the natural sciences.

Yet, although there are basic similarities between faunas, there are also great differences. If you blindfolded a party of naturalists, took them on a mystery plane trip and dumped them in a desert, they should be able to tell you fairly quickly on what continent they had landed. If the plants around them had bloated, green stems, lacked leaves, and were covered with spines, then they would know that they were in the Americas, for cacti only occur naturally in the New World. If they spotted a large animal with a baby peering out from a pouch on its stomach, they would know they were in Australia. And if a herd of large, hoofed, striped animals galloped away from them into the distance, they would be certain they were in Africa. How is it, then, that in spite of the inter-relatedness of animals worldwide, different continents have their own special versions?

Variations on a theme

The answer to that conundrum was provided in the middle of the last century by Charles Darwin. He demonstrated that species of animals do not remain fixed in their characters for all time but, over many generations, change, so that one species gives rise to others and, over extremely long periods of time, the accumulated changes produce quite new kinds of animals. He called this process evolution by natural selection. Animal populations on the supercontinent were evolving both before and after its breakup, and since all its fragments did not separate simultaneously, different sections carried away different mixes of animals. Those populations, isolated on their continents, continue to evolve, eventually producing their own characteristic variations on the near-universal themes of frogs and spiders, butterflies and snakes.

So for a full understanding of the present state of the natural world, we have to look at its history. That history is not only unimaginably long but full of great changes. During the 4 billion years or so since life first appeared, the planet has warmed and cooled and continents have drifted apart and collided. There

have been periods of violent activity worldwide and others when the globe has been comparatively quiescent. And throughout that vast expanse of time, living organisms, both plant and animal, steadily evolved into more and more complex forms and spread to almost every part of the Earth and sea. Sometimes vast catastrophes, the exact character of which is still a mystery to us, annihilated tens of thousands of species, but such mass extinctions only led to an even greater burgeoning of new forms among the survivors.

Evolution is still active

Are such gigantic changes only things of the past? Viewed with the perspective of our own brief lives, the world today seems such a stable place that talk of continents colliding and new categories of animals suddenly appearing seems a little farfetched. But once we have recognized these processes in the past, we know what trends to look for in the present. Observational satellites in space enable us to check whether or not the continents are still drifting. They are. The width of the Atlantic Ocean has now been measured many times with unprecedented accuracy. It is steadily increasing, and North America and Europe are now moving apart at the rate of an inch or so a year. The great continental blocks of India and central Asia, which first collided about 60 million years ago, are still moving, compacting with one another and pushing up the buckled sedimentary rocks of the Himalayas still higher.

Changes in the shapes of animals' bodies are not so easily observed. Since alternations in genetic makeup can only occur as each new generation is conceived, it follows that we cannot witness in our own lifetimes the cumulative effect of many generational changes in animals whose longevity is about the same as ours. Not for us the privilege of knowing whether and how elephants, which live even longer than we do, are in the process of evolving into new forms. If we want to see such changes taking place, we must look at animals such as aphids and fruit flies that can produce several hundred generations in a few years. And when we do that, we can indeed confirm that evolution is still active in the world.

The realization that life on Earth is even now in the process of changing may well lead us to speculate about its future. But more important than that, understanding those processes could enable us to safeguard its survival and its destiny.

Origins of our planet

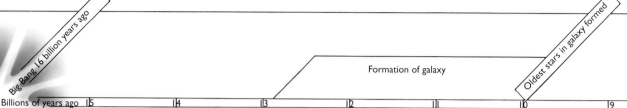

Big Bang 16 billion years ago

Formation of galaxy

Oldest stars in galaxy formed

Billions of years ago | 15 | 14 | 13 | 12 | 11 | 10 | 9

The only life in the universe of which we have evidence began on this Earth around four billion years ago. The conditions present on its surface at that time provided the basic ingredients—liquid water and carbon compounds—out of which all known life is constructed. But where did the Earth come from? Why were the conditions appropriate for the creation of life to be found here and, so far as is known, nowhere else in the solar system?

All matter in the universe was originally formed during the early phases of the "Big Bang" that brought the universe into existence—an explosion of unimaginable violence which took place about 16 billion years ago. Beginning as a small, intensely hot entity, the universe has from that moment on been expanding and cooling. In the early stages of its expansion, it contained only subatomic particles and waves of radiation. Later, the nuclei of hydrogen atoms coalesced. Subsequently, fusion of these basic nuclei and their capture of electrons led to the creation of all the other, heavier elements.

The matter out of which the early universe was composed was not distributed evenly, however, and gravitational attraction brought about further clusterings. Such giant concentrations of gas and dust were the starting material for galaxies— island subuniverses each containing enough matter for billions of stars. In each galactic mass, individual clouds of matter were the beginnings of stars like our sun.

Birth of the solar system

Modern theories vary, but most agree that the sun did not "capture" its family of circling planets by gravitational attraction. They suggest that the entire solar system was created at the same time, between 4.5 and 5 billion years ago. The story probably began with a roughly spherical cloud of hydrogen and dust. Gravity caused this cloud to collapse toward its own center and begin to rotate. But as it picked up speed, the shape of the giant spinning cloud began to change. It flattened out like a disk, with a spherical central bulge.

Hydrogen nuclei at the center of the bulge were squeezed by gravity and reached such a high density that fusion— the process that fuels the hydrogen bomb—began to occur. Helium nuclei were produced and enormous quantities of energy were given off in the form of heat and light. The sun had begun to shine.

Elsewhere in the spinning disk, most of the swirling gas and dust particles collided and stuck together, "glued" by electrostatic and other forces to form ever-larger "rocks." Eventually these acquired sufficient mass to exercise a strong gravitational attraction, and planets were formed from the aggregated matter.

The four planets that lie closest to the heat of the sun—Mercury, Venus, Earth, and Mars—were formed from clusters of material with high melting points: all are rocky, with metal cores. Farther out are the frigid gas giants: Jupiter, Saturn, Uranus, and Neptune.

Early on in its history, the Earth's substance sorted itself into layers. Heated to over 7,000°F (4,000°C) by the natural

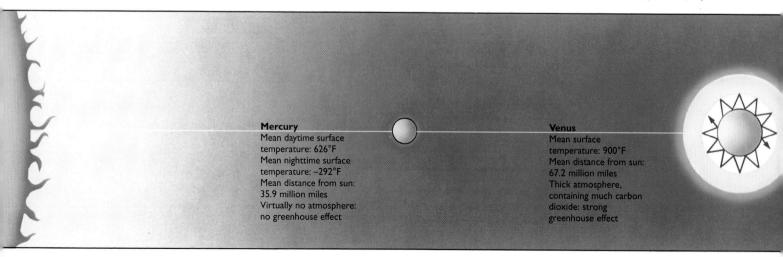

Mercury
Mean daytime surface temperature: 626°F
Mean nighttime surface temperature: −292°F
Mean distance from sun: 35.9 million miles
Virtually no atmosphere: no greenhouse effect

Venus
Mean surface temperature: 900°F
Mean distance from sun: 67.2 million miles
Thick atmosphere, containing much carbon dioxide: strong greenhouse effect

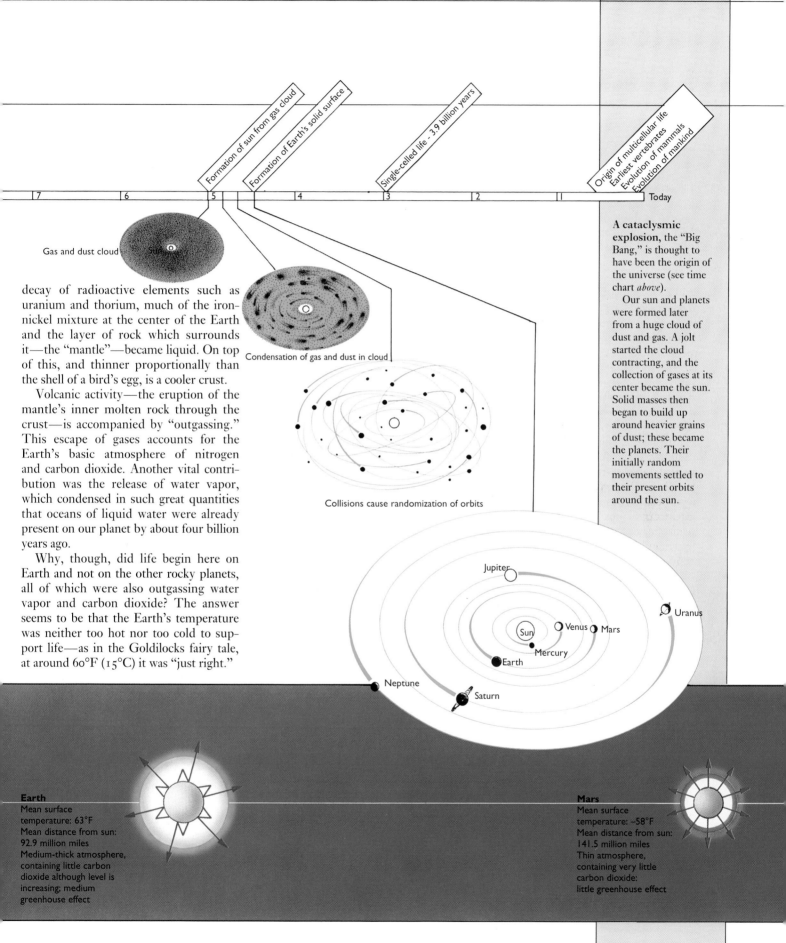

Formation of sun from gas cloud

Formation of Earth's solid surface

Single-celled life - 3.9 billion years

Origin of multicellular life
Earliest vertebrates
Evolution of mammals
Evolution of mankind

7 6 5 4 3 2 1 Today

Gas and dust cloud

Sun

Condensation of gas and dust in cloud

Collisions cause randomization of orbits

decay of radioactive elements such as uranium and thorium, much of the iron-nickel mixture at the center of the Earth and the layer of rock which surrounds it—the "mantle"—became liquid. On top of this, and thinner proportionally than the shell of a bird's egg, is a cooler crust.

Volcanic activity—the eruption of the mantle's inner molten rock through the crust—is accompanied by "outgassing." This escape of gases accounts for the Earth's basic atmosphere of nitrogen and carbon dioxide. Another vital contribution was the release of water vapor, which condensed in such great quantities that oceans of liquid water were already present on our planet by about four billion years ago.

Why, though, did life begin here on Earth and not on the other rocky planets, all of which were also outgassing water vapor and carbon dioxide? The answer seems to be that the Earth's temperature was neither too hot nor too cold to support life—as in the Goldilocks fairy tale, at around 60°F (15°C) it was "just right."

A cataclysmic explosion, the "Big Bang," is thought to have been the origin of the universe (see time chart *above*).

Our sun and planets were formed later from a huge cloud of dust and gas. A jolt started the cloud contracting, and the collection of gases at its center became the sun. Solid masses then began to build up around heavier grains of dust; these became the planets. Their initially random movements settled to their present orbits around the sun.

Jupiter

Sun Venus Mars

Uranus

Mercury

Earth

Neptune

Saturn

Earth
Mean surface temperature: 63°F
Mean distance from sun: 92.9 million miles
Medium-thick atmosphere, containing little carbon dioxide although level is increasing; medium greenhouse effect

Mars
Mean surface temperature: −58°F
Mean distance from sun: 141.5 million miles
Thin atmosphere, containing very little carbon dioxide: little greenhouse effect

The moving face of Earth

The appearances of the world are deceptive. The land, with its ranges of mountains raising their apparently ancient snowy peaks into the skies, seems massive and unmoving. Yet, in reality, every aspect of the world is changing, however minutely, day by day and month by month.

The seemingly stable continents are in fact slowly moving and, perhaps, starting to fragment. America and Europe are gradually moving farther apart. Some of those ancient mountains are in truth young and are still rising skyward.

The energy for these great changes comes from within the Earth. As the molten lavas that pour from volcanoes so dramatically illustrate, the depths of the Earth are hot—so hot that the rocks themselves are liquid. The source of this heat is the radioactivity of many elements, particularly of some isotopes of uranium, thorium, and potassium. Just as the water heated at the bottom of a kettle rises and circulates, so the heat produced by the radioactive decay of these elements causes patterns of movement of the rocks themselves.

The structure of the Earth

These patterns do not involve all the material of the Earth, which has a layered structure. As it cooled from its original molten state, the heaviest elements and compounds sank to the center of the planet. There they formed a dense core. Nearly all of the remainder of the Earth is made up of the "mantle," which is about 1,800 miles deep. On top of this, like the thin skin of a peach, lies the "crust," which is made up of the lightest minerals.

The continents themselves contain many radioactive minerals, and are therefore comparatively hot. The heat that rises from deeper within the mantle is mainly lost through the thinner crust that lies below the oceans. It is here that upwelling heated material rises to the surface, where it must then move laterally before it can descend, now cooled, into the Earth. Where it reaches the surface, great submarine mountain chains, the "spreading

The Earth's plates

Eurasian Plate
American Plate
Philippine Plate
Caribbean Plate
African Plate
Pacific Plate
Indian-Australian Plate
Antarctic Plate

Trench
Oceanic ridge

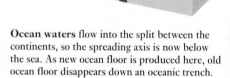

Oceanic floor
Continent

Heated material from within the Earth's mantle rises to the surface below a landmass. This material spreads out and splits the landmass into two smaller continents.

Ocean waters flow into the split between the continents, so the spreading axis is now below the sea. As new ocean floor is produced here, old ocean floor disappears down an oceanic trench.

ridges," form in the oceans. Here, new ocean floor is continually being produced as the regions on either side move apart. This movement is slow, only an inch or so a year—about the same rate at which our fingernails grow—but even such a tiny amount as this gradually builds up over the centuries.

The cooled crust descends into the Earth at a system of great troughs or trenches, about 6 miles deep and 155 –185 miles wide, many of which ring the Pacific Ocean. From there, the cooled material moves downward to a depth of about 125 miles, where it melts. Later, it may rise again to a new spreading center

and form new ocean floor. The ocean floor is therefore being slowly but continually cycled and is comparatively youthful—none of it is more than 200 million years old. The ancient rocks of the continents are up to 4 billion years old.

Evidence for this process of ocean floor spreading is provided by the clear pattern of the age of the floor: it is always youngest adjacent to the spreading ridges, where it is formed, and oldest near the trenches, where it disappears into the Earth.

The surface of the Earth is divided into a pattern of "plates," which move in different directions away from the spreading ridges and toward the trenches. Some, but

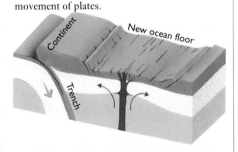

The major plates into which the surface of Earth is divided are shown on the map. These plates are slowly moving in different directions, away from the spreading oceanic ridges and toward the trenches. Arrows show the direction of movement of plates.

The Earth is made up of three major "layers"—core, mantle, and crust. The crust is 3 miles thick below the oceans but 20–30 miles thick below the continents.

All the ocean floor of the westward-moving continent has disappeared, but the continent cannot follow it. Instead, ocean floor of the adjacent plate is now being consumed.

not all, of these plates bear continents, and so the whole system of movement is now known as "plate tectonics" rather than continental drift.

Perhaps the most dramatic evidence for the occurrence of plate tectonics lies in the phenomenon of paleomagnetism. Many rocks contain magnetized particles that become aligned with the Earth's magnetic field as they are deposited in the rocks.

These tiny fossil magnets reveal where those rocks were in relation to the magnetic poles when they were laid down. By studying a series of rocks from one continent, its movement can be tracked relative to the magnetic pole. For example, North

Africa lay over the south magnetic pole some 400 million years ago and has slowly drifted to its present position.

By comparing the tracks of different continents, we can see how they have moved relative to one another. This can show how and when they collided with one another or split apart. By moving them on the map back down their tracks, we can see how they were once positioned. The fact that this process leads to a pattern that is consistent with the shapes of the edges of the continents, and with the patterns of ages of the rocks, is overwhelming evidence of the correctness of the theory of plate tectonics.

The Himalaya Mountains were created by geological processes that began around 150 million years ago when a giant "raft" of the Earth's crust, bearing what is now India, broke away from the huge southern landmass. The raft drifted northward and collided 60 million years ago with Asia. The intervening former ocean floor was buckled into a vast mountain range.

The origins of life

Life is a comparatively fragile phenomenon that can exist only within a narrow range of conditions. Yet in the vastness of the universe, it is assumed that the conditions necessary for the evolution of life must have occurred many times.

The Earth on which life first evolved was, however, very different from the planet of today: the very oxygen that we breathe, and the protective ozone layer of the Earth's atmosphere, did not exist then, for they are themselves the products of that life.

The oldest direct traces of life on Earth date back almost 4 billion years. In rocks that age in Australia and southern Africa, geologists have found stromatolites, layered structures created through the activity of primitive bacteria. Other Australian rocks of similar age provide even more direct evidence of ancient life. Sections of these rocks, known as cherts, show the fossilized remains of blue-green bacteria themselves.

Rocks also reveal even more distant, indirect traces of life. Living things use particular isotopes (physical forms) of the element carbon preferentially. The mix of carbon isotopes detected in rocks from Greenland more than 3.8 billion years old also show evidence of life on Earth—that is, only 600 million years after the planet itself was first formed.

What is life?

Reduced to its barest essentials, life is the ability of an organic substance to produce a replica of itself. In organisms alive today, that ability is found only in molecules like DNA—deoxyribonucleic acid.

The huge information-carrying molecule of DNA is similar to a set of detailed plans, and the plans are for making proteins. From proteins, whose production and packaging is controlled by the endoplasmic reticulum and Golgi bodies of plant and animal cells, DNA can build itself a cell. When the cell divides in two, each daughter cell needs a copy of the master plan for making new cells. So, before the division, the DNA copies itself so that a version of the DNA plan can be passed into each of the two newly formed cells.

In the blue-green bacteria, organisms with ancient origins, the DNA is found loose in an uncomplicated cell. This "prokaryote" structure is simpler than the "eukaryote" one, in which the DNA is organized into threadlike chromosomes and housed within a cell nucleus. Eukaryotes are thought to have evolved about 1.2 billion years ago.

Whether simple or complex, all living matter is made primarily of the compounds of carbon, oxygen, hydrogen, and nitrogen, which would have been abundant in the Earth's early atmosphere in the form of gases such as water vapor, nitrogen, and carbon dioxide. Each of these gases is given off during volcanic activity, which was very frequent at that time.

If ultraviolet light or lightning activates such an atmosphere, a variety of simple organic compounds are formed. An early form of DNA could have been made from such ingredients. Once the first DNA-like molecule had appeared, it would rapidly have reproduced itself. Then natural selection would have come into play, favoring the survival of those variants best adapted to quick, effective reproduction and vigorous offspring.

DNA replication became much more effective through the evolution of the cell, within whose walls or membranes are substances such as proteins. Moreover, amino acids, the building blocks from which proteins are constructed, are spontaneously formed in the chemically reactive conditions believed to be present in the Earth's early history.

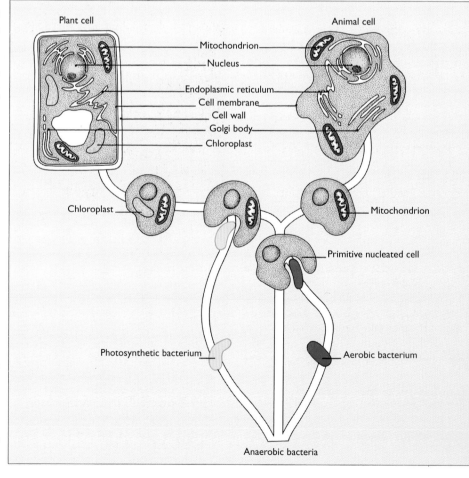

Another development of immense importance was the evolution of the green pigment chlorophyll, for this allowed cells to trap sunlight and use it to produce their own energy. As a by-product of this process of photosynthesis, early photosynthetic bacteria, like their modern higher plant counterparts, released oxygen.

The oxygen produced by plants at first became "locked up" by reacting or combining with other substances and minerals. Eventually, some 2.2 billion years ago, free oxygen was present in the atmosphere. Living things used this reactive substance in the biochemical functions of their own cells. The free oxygen in the atmosphere also produced a layer of ozone, which filters out the ultraviolet light from the sun that is harmful to life below.

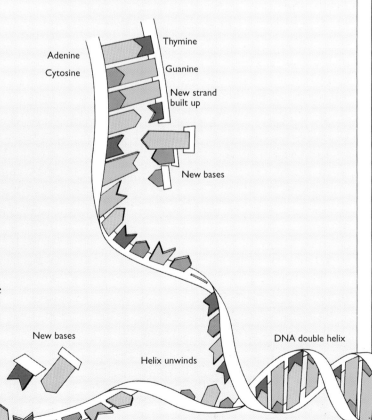

Complex cells (*left*) may have evolved by the permanent combination of more simple cells. If a simple anaerobic bacterium, an organism living without oxygen, engulfed an aerobic bacterium, the newcomer could have become a mitochondrion, the cell organelle that uses oxygen to provide energy. In a similar way, the chlorophyll-containing chloroplasts typical of plant cells may have originated as photosynthetic bacteria. Plant cells also differ from those of animals in having thick, rigid walls.

The DNA molecule (*right*) is at the heart of all earthly life and is built like a ladder. Each "rung" is made of a pair of chemicals ("bases")—either adenine and rhymine, or cytosine and guanine. When the molecule replicates itself, the helix unwinds and new bases are added from a "pool" available in the cell. Thus two perfect copies of the original helix are produced.

Adenine
Cytosine
Thymine
Guanine
New strand built up
New bases
New bases
New strand built up
Helix unwinds
DNA double helix

Some of the oldest evidence of life on Earth, dating back about 3.5 billion years, is contained in stromatolites, rocklike structures such as these at Shark Bay, Australia.

They are formed by primitive organisms, blue-green bacteria. As the bacteria grow they form a web of material in which sediments from the surrounding water are trapped and eventually compacted into a dense mat.

Another bacterial web then forms above this layer, and this too forms a sediment mat. The process continues, gradually shaping the mound- or even pillar-like structures known as stromatolites.

The distant past

Over many millions of years, the simple form of life that had begun on our planet gradually became more complex. At first living organisms remained in the sea, where they had evolved. But, some 500 million years ago, in the Ordovician period (see Geological Time chart, p.218), life forms extended into freshwaters and eventually onto the land.

The first land plants date from about 500 million years ago. They were probably accompanied by the first terrestrial invertebrates, such as arthropods and worms, that fed on the decaying plant matter.

Vertebrate animals did not colonize the land until the Devonian, about 380 million years ago, but soon evolved into a variety of amphibians and reptiles. Their distribution provides the first opportunity to analyze the biogeography of land organisms. Such studies have revealed, for example, that land vertebrates seem to have evolved in what was the continent of Euramerica. Nearly all the earliest evidence of their existence, from the Late Devonian until the middle of the Permian period, has been found there.

At this time Euramerica lay across the equator. Large parts were covered by swampy tropical forest, similar to that of the Amazon jungle today. These forests contained early relatives of the conifers, club mosses, and horsetails, as well as tree ferns and seed ferns.

Farther north, on the separate continent that eventually formed what is now Siberia, there was a different flora, adapted to a cooler climate, with many conifer-like trees and seed ferns. The areas

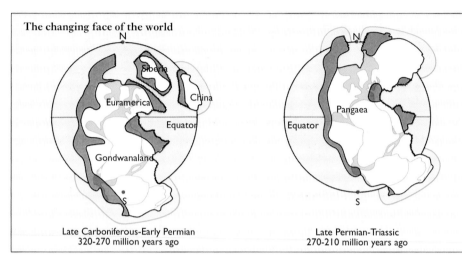

The changing face of the world

Late Carboniferous-Early Permian
320-270 million years ago

Late Permian-Triassic
270-210 million years ago

destined to become China also had distinct floras, and the southern landmass of Gondwanaland had a unique flora of cold-adapted plants that grew to within 5 degrees of latitude of the South Pole. Biologists had already identified these separate floras long before geologists realized that they lay on separate continents.

The early vertebrates

Although the great southern landmass of Gondwanaland is thought to have joined up with Euramerica in the Late Carboniferous, there is no evidence that the land vertebrates of Euramerica dispersed southward into it at this time. A mountain barrier that rose as the two continents collided is the most likely reason for this. The now-eroded relics of those mountains remain in the Allegheny Mountains of the eastern United States and the Atlas range of northwest Africa.

Fossil records suggest that it was not until 265 million years ago that land vertebrates were able to cross this mountain barrier and disperse into Gondwanaland. And it was at that time that Siberia attached itself to what was now a world-continent, Pangaea, over which land vertebrates roamed freely. In the Triassic, for example, around 230 million years ago, similar faunas, both amphibians and

Cynognathus was a mammal-like reptile, the group from which the first mammals evolved.

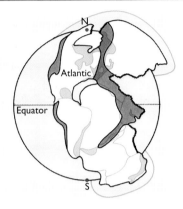 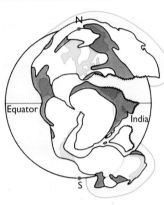

The four globes show how the pattern of the world's landmasses changed over a period of about 300 million years. Areas at the "back" of the globe have been folded out. Dotted lines indicate the coastlines of modern continents.

■ Shallow seas
□ Oceans
〰 Continental edges that later collide

Mid-Jurassic
180 million years ago

Early Cretaceous
100 million years ago

About 300 million years ago much of Euramerica was covered with tropical swamp forests, filled with giant trees.

Peat accumulation in these swamps led to the development of extensive domes similar to modern raised bogs.

In some areas the land covered by these forests was slowly sinking, forming basins that filled with the accumulated remains of the trees. These became compressed, dried, and hardened to form the coal deposits of the eastern United States, Britain, and central Europe.

reptiles, were found in the lands that are now Australia, southern Africa, North and South America, Europe, and Asia.

The Triassic fauna of Pangaea contained a variety of groups of amphibians and reptiles. And it was in this Triassic world that two major evolutionary events took place. One great group of reptiles, the mammal-like reptiles, evolved into the mammals. At almost the same time, another group, the archosaurs, evolved into the dinosaurs. Both groups were therefore able to spread throughout Pangaea.

If, however, these groups had evolved later, when the world-continent had split up into separate landmasses, the outcome might have been very different. If some continents had contained only mammals and others only dinosaurs, the two groups might have evolved independently. Instead of dinosaurs becoming only gigantic uninsulated creatures that ruled the daylight hours, some of them might have become small, warmly covered animals. Similarly, instead of becoming only small, furry denizens of the night, mammals might have evolved into larger creatures much earlier.

As it was, in the presence of the mighty dinosaurs, mammals had to remain insignificant nocturnal creatures until they were able to grasp the great new opportunity when dinosaurs became extinct. In this way evolution was influenced by geography.

Ornithosuchus was an archosaurian reptile. Dinosaurs evolved from forms similar to *Ornithosuchus* and spread throughout the world.

Hypsilophodon was a dinosaur that occurred in both Europe and North America. It lived in the Early Cretaceous period when those areas were still linked, so it could roam widely.

19

The last
100 million years

About 100 million years ago, at the beginning of the Late Cretaceous period, the two landmasses in the northern hemisphere, although still joined together, had each been bisected by a seaway. Extending south from the Arctic Ocean, the Interior Seaway ran down North America to the Gulf of Mexico, while the Tugai Sea separated Europe from Asia (see map). The resulting landmasses were Asiamerica, composed of Asia plus western North America, and Euramerica, made up of eastern North America plus Europe. This Cretaceous geography, deduced by geologists, is confirmed by the distribution of the living creatures of the Late Cretaceous.

All the Cretaceous continents were populated by the older types of dinosaur that had evolved during the Jurassic period. Similarly, the ostrich-dinosaurs, the dome-headed dinosaurs and early duck-billed dinosaurs, which evolved in Early Cretaceous times before the seaways appeared, had spread throughout the northern hemisphere. But the ancestors of the horned ceratopsian dinosaurs, as well as the massive carnivorous tyrannosaurs and the more advanced types of duck-billed dinosaur, all evolved in Asia after the appearance of the seaways. So they were able to move from Asia, via Alaska, into western North America. Some also made their way, perhaps via a chain of islands, into South America; but, confined by the seaways to east and west, they were never able to colonize Euramerica.

Placental mammals, too, evolved in Asia and soon spread to western North America. The pouched mammals, the marsupials, evolved somewhere in the linked chain of southern continents—South America—Antarctica—Australia. Though some marsupials dispersed northward to North America, most eventually became extinct there in the face of competition from placental mammals.

At the end of the Cretaceous, around 70 million years ago, the Interior Seaway gradually withdrew from North America, reuniting it with Europe in a new, though short-lived, Early Cenozoic Euramerica, which seems to have been the main theater for the radiation of the placental mammals. The first hoofed mammals (both the split-toed and the horselike forms), the first primates, bats, rodents,

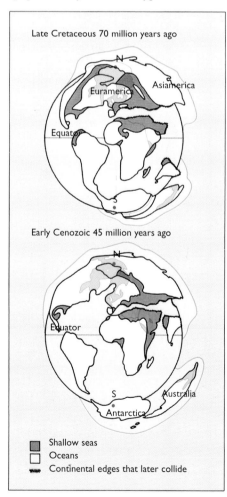

Late Cretaceous 70 million years ago

Asiamerica
Euramerica
Equator

Early Cenozoic 45 million years ago

Equator
Australia
Antarctica

■ Shallow seas
□ Oceans
〰 Continental edges that later collide

The Isle of Sheppey, near London, England

insectivores, and carnivores all originated in that era. Not until the Eocene epoch, about 50 million years ago, did Greenland separate from Europe to break the land link between North America and Europe.

The spread of plants

Flowering plants seem to have evolved by the Early Cretaceous (135 million years ago) in the tropical, near-equatorial region of the still-linked continents of South America and Africa. Aided by their light seeds, which can be wind-borne, they spread rapidly around the world and adapted to its various climates. The tropical types spread through Africa and into southern parts of Europe, Asia, and North America, and elements of this flora must have adapted to the cooler climates of

northern North America and Eurasia.

In the southern hemisphere it is possible to distinguish a very different flora adapted to the cooler temperate climate. Characterized by the southern beech tree, *Nothofagus*, this flora spread from southern South America across the Antarctic to Australia and New Zealand. This was possible because in these Early Cenozoic days the polar regions were still unglaciated, and milder climates extended much farther toward the poles.

So, for example, 50 million years ago a rich fauna of salamanders, turtles, tortoises, lizards, snakes, alligators, tapirs, and flying lemurs lived on Ellesmere Island in the Canadian Arctic. And a subtropical flora and fauna flourished in southeastern England.

About 50 million years ago a rich subtropical flora grew on the Isle of Sheppey. The land was fringed with mangrove trees, many types of palm, and relatives of the living magnolia, cinnamon tree, dogwood, and grape. Lianas (climbers) festooned the trees and ferns, sedges, and club moss carpeted the ground; grasses had not evolved at this time.

The two globes show the pattern of changes in the world's landmasses over the last 100 million years. Areas at the "back" of the globe have been folded out. Dotted lines indicate the coastlines of modern continents.
1 *Meliosoma*
2 *Cinnamomum* 3 *Platycarya*
4 *Mastixia* 5 *Menispermum*
6 *Uvaria* 7 *Nipa* 8 *Vitis*
9 *Magnolia*

Southeast England
London
Isle of Sheppey

The Isle of Sheppey lies just off the coast of southeast England near the Thames estuary. The landscape in today's cooler climate (below) is very different from that of 50 million years ago.

Patterns of today

The living world of today can be conveniently divided into regions, each with its characteristic mammals and flowering plants. Every such region has some unique types. For example, the kangaroo, sloth, and giraffe are unique to the continents of Australia, South America, and Africa respectively. As a rule continents that have been isolated for a long time have many unique organisms; those that have always had land connections with others have few.

The southern continents

Australia, with its many marsupials, has the most distinctive fauna of all. It has been isolated for at least 50 million years, longer than any other continent, and few of the placental mammals that dominate the rest of the world ever reached it. Although its neighbor Antarctica may once have had a similar fauna, the ice sheets that now cover that continent make present-day comparisons impossible.

South America was also an isolated continent until about three million years ago and was home to a mixture of marsupials and primitive placental mammals. Most of this distinctive fauna became extinct when the formation of the Panama land bridge between North and South America allowed more advanced placental mammals to enter from the north (see pp. 26–27). The anteaters, armadillos, sloths, New World monkeys, and marsupial opossums are now the only relics of the rich original fauna of South America.

Africa, although physically united with the lands to the north, has never been easy for animals to reach. In the Tertiary era the Mediterranean Sea was wider than it is today, and only a few types of mammal reached Africa at that time. Later, the great deserts of northern Africa and the Middle East continued that early isolation. Some of the first mammals to enter the continent evolved into characteristically African groups—elephants, conies, elephant shrews, and the aardvark.

India and Southeast Asia are recognized by zoologists as being a separate region—the Oriental. The flora of this area is less distinct, but the eastern "Indomalesian" region can be separated from an African "Ethiopian" region. Little is known of India's early fauna and flora when, as a separate island continent, it drifted through the Indian Ocean until it collided with Asia.

In the Early Miocene the shallow seas that had separated India from Africa withdrew, allowing an exchange of mammals. Later the two regions were again separated by the growing deserts of the Middle East. Hence, although both regions contain elephants, rhinos, apes, and smaller primates, those of the Oriental region have evolved separately and are different from those of Africa.

The northern continents

The great landmasses of the northern hemisphere, North America and Eurasia, now have temperate climates and few species of animals and plants compared with lands to the south. But until nearly two million years ago, the faunas and floras of Africa and India extended northward into Eurasia. Hippos, apes, rhinos, giraffes, tapirs, hyenas, and elephants all occurred there, and it was the ice ages that led to their extinction in the north.

North America did not receive such tropical animals, partly because the southern part of the continent was much drier than their lush tropical homelands of South America, and partly because the Panama land bridge, the sole link with

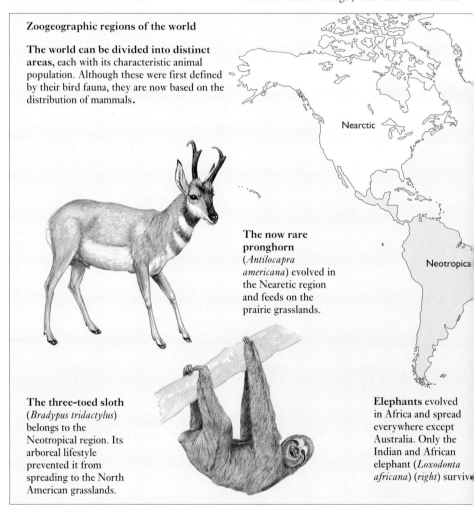

Zoogeographic regions of the world

The world can be divided into distinct areas, each with its characteristic animal population. Although these were first defined by their bird fauna, they are now based on the distribution of mammals.

Nearctic

Neotropica

The now rare pronghorn (*Antilocapra americana*) evolved in the Nearctic region and feeds on the prairie grasslands.

The three-toed sloth (*Bradypus tridactylus*) belongs to the Neotropical region. Its arboreal lifestyle prevented it from spreading to the North American grasslands.

Elephants evolved in Africa and spread everywhere except Australia. Only the Indian and African elephant (*Loxodonta africana*) (*right*) survive

South America, was only completed three million years ago. Until then immigrants could only come via the Bering land bridge that joined Alaska and Siberia until a few thousand years ago. This northerly area was very vulnerable to the climatic changes of ice ages and, as the climate cooled, it was only the larger, hardy animals that could make the journey. Thus the elk, moose, and caribou (known as the red deer, elk, and reindeer in Europe) are found in both Europe and North America as are bison and musk ox.

The northern continents have few unique mammals. Only the pronghorn antelope is unique to North America while Eurasia has only a single unique rodent family.

Floral regions of the world

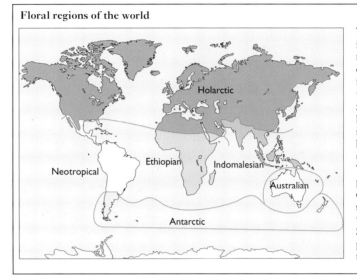

The world divides into six floral regions based on the distribution of flowering plants. But, whereas each zoogeographic region has characteristic inhabitants, it is far harder to pinpoint unique plant examples. This is partly because plants are more easily dispersed than animals, so tend to be more widely distributed. Some 86 of the 302 plant families are found worldwide.

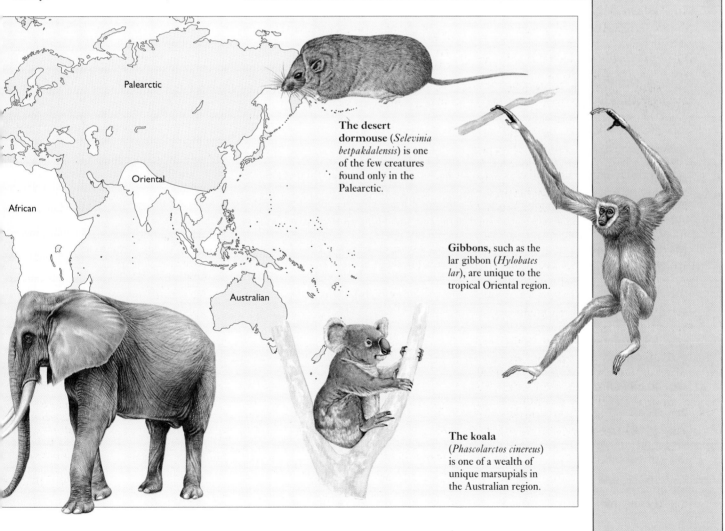

The desert dormouse (*Selevinia betpakdalensis*) is one of the few creatures found only in the Palearctic.

Gibbons, such as the lar gibbon (*Hylobates lar*), are unique to the tropical Oriental region.

The koala (*Phascolarctos cinereus*) is one of a wealth of unique marsupials in the Australian region.

23

The isolated continent

Australia stands apart from the other continents of the world in its living and nonliving features. Its uniqueness results from several factors: its isolation comparatively early in the Earth's history, so that many late-evolving types of animals and plants were unable to reach it; its mainly flat terrain, with few mountains; and its climatic history, arising from its movement from a cool-temperate to an arid latitude.

Australia was originally a part of the great southern landmass, Gondwanaland (see pp. 18–19). First India, and later Africa, split off from the landmass, leaving Australia and South America still interconnected by Antarctica. South America split away about 70 million years ago, leaving Australia and Antarctica. At first they moved northward, so that the latitude of central Australia changed from about 55° south to about 45° south. At that time Australia had a warm climate, with an adequate rainfall from the westerly winds that circled the southern end of the world. The continent continued to move northward until 55 million years ago, after which the gulf between Australia and Antarctica widened more rapidly

Knowledge of these movements helps to explain how and why Australia came to possess its unusual fauna and flora.

Flowering plants probably spread to Australia from the equatorial regions, where they evolved, about 90 million years ago, when the southern continents were still connected. Living representatives of these plants still flourish in a few small surviving areas of rain forest. But most of the country now has a very different flora, resulting from its geological and climatic history.

Australia is the flattest of all continents. Its only mountains lie along its eastern margin, and these are old (about 200 million years) and heavily eroded (less than 6,550 ft high). Because there are no new, young mountains to be weathered away, the great plains of Australia's heartland have received no new sediments. Their soils are old, highly weathered, poor in nutrients and minerals.

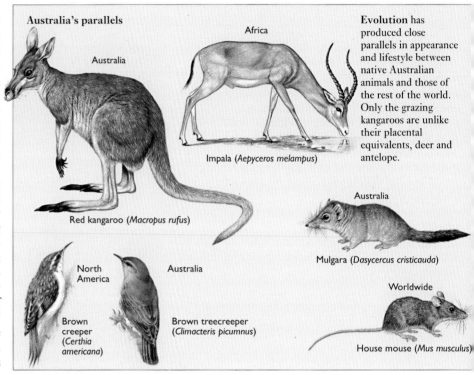

Australia's parallels

Red kangaroo (*Macropus rufus*)

Africa

Australia

Impala (*Aepyceros melampus*)

Evolution has produced close parallels in appearance and lifestyle between native Australian animals and those of the rest of the world. Only the grazing kangaroos are unlike their placental equivalents, deer and antelope.

Australia

Mulgara (*Dasycercus cristicauda*)

North America

Australia

Brown creeper (*Certhia americana*)

Brown treecreeper (*Climacteris picumnus*)

Worldwide

House mouse (*Mus musculus*)

The plants that have colonized this difficult environment are particularly hardy, and are described as sclerophylls. They have sparse, small evergreen leaves and can stop growing temporarily if conditions become impossible; the most common examples belong to the genera *Eucalyptus* and *Acacia*. Although so different from rain forest plants, sclerophylls have clearly evolved from them, since most have close relatives in that flora.

Even sclerophyll plants, however, are now found only in scattered areas around the periphery of Australia, because climatic history made the environment even more difficult for plants. As Australia separated from Antarctica and moved north, a new cold, deepwater ocean current developed, sweeping between the two continents, whose climates now diverged.

Antarctica became progressively colder, as its glaciers formed and grew. Little water condensed from the chill Antarctic seas into rain clouds to moisten the vast flat lands of Australia. And as that continent drifted farther and farther north, it moved into the 30° south low pressure

area where there is little rainfall. Australia became the driest continent on Earth, two-thirds receiving under 20 in of rain per year and the remainder less than 10 in. As a result, desert and grass shrublands cover most of Australia today.

Australian fauna

By chance, an unusual group of mammals was given the opportunity to evolve in this unique environment. The placental mammals, which originated in Asia and spread to the rest of the world, did not reach Australia. The pouched marsupial mammals, however, which seem to have evolved in North America, did spread beyond South America into Antarctica and Australia. They were then free to fill all the different ecological niches in Australia, unhampered by competition from their placental relatives.

The two groups closely parallel each other. The marsupial versions of wolf, cat, mongoose, mouse, anteater, marmot, bear, squirrel, and so on, often look remarkably similar to their placental counterparts.

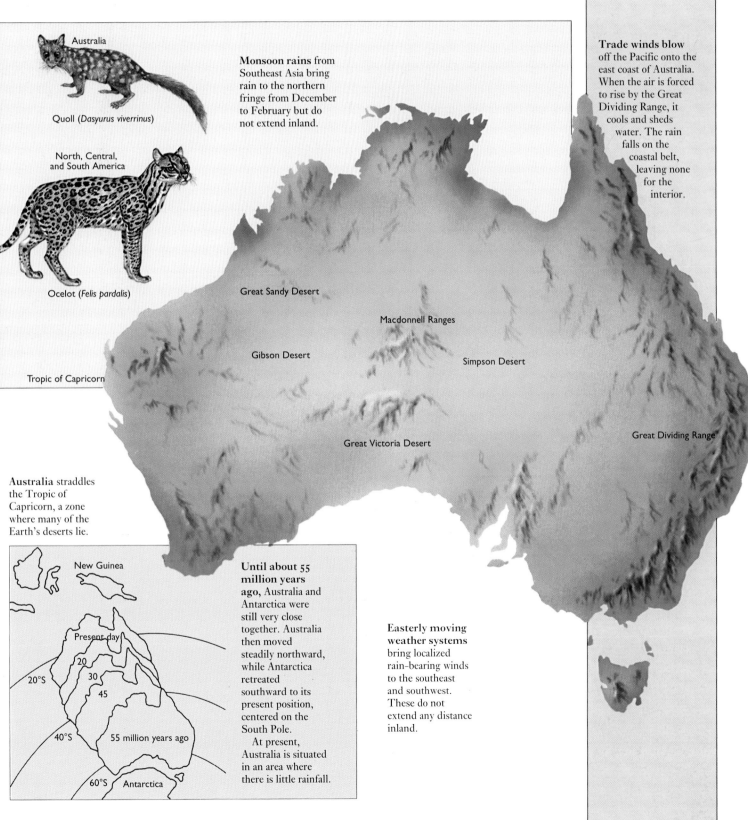

Australia

Quoll (*Dasyurus viverrinus*)

North, Central, and South America

Ocelot (*Felis pardalis*)

Monsoon rains from Southeast Asia bring rain to the northern fringe from December to February but do not extend inland.

Trade winds blow off the Pacific onto the east coast of Australia. When the air is forced to rise by the Great Dividing Range, it cools and sheds water. The rain falls on the coastal belt, leaving none for the interior.

Great Sandy Desert

Macdonnell Ranges

Gibson Desert

Simpson Desert

Tropic of Capricorn

Great Victoria Desert

Great Dividing Range

Australia straddles the Tropic of Capricorn, a zone where many of the Earth's deserts lie.

New Guinea

Present day

20

20°S

30

45

40°S

55 million years ago

60°S   Antarctica

Until about 55 million years ago, Australia and Antarctica were still very close together. Australia then moved steadily northward, while Antarctica retreated southward to its present position, centered on the South Pole.

At present, Australia is situated in an area where there is little rainfall.

Easterly moving weather systems bring localized rain-bearing winds to the southeast and southwest. These do not extend any distance inland.

25

The great American interchange

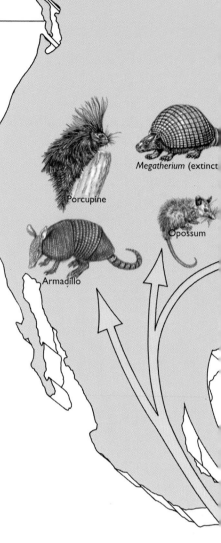

Megatherium (extinct

Porcupine

Armadillo

Opossum

No other continent has experienced such changes in its fauna as South America. The basic cause of these changes was the movement of tectonic plates, which rafted the continent away from its original partner, Africa, and eventually linked it instead to North America. During the periods when it was linked to other landmasses, South America received immigrants from them; when it again became isolated, its new fauna had the opportunity to evolve and diversify, until a new connection brought a further period of immigration and integration.

Island-hopping

Near the beginning of the Tertiary period, (around 60 million years ago) some North American marsupial and placental mammals managed to disperse to South America, probably by a rather hazardous island-hopping route such as that provided by the Caribbean island chain today. Where marsupial and placental mammals have coexisted elsewhere, the placentals have normally dominated the marsupials, which have gradually diminished in variety. But in South America, the marsupials became quite diverse and evolved into types similar to the mountain lion, mole, and kangaroo-rat, as well as into insectivorous forms.

Early South American placental mammals evolved into a variety of groups. One of these, the edentates, which includes the armadillos, anteaters, and tree sloths, still exists, although the early sloths were elephant-sized terrestrial animals, far larger than their arboreal descendants. Another group, now extinct, evolved into a variety of strange, hoofed herbivores, which paralleled the horses, camels, tapirs, and elephants that developed in the rest of the world.

In the Oligocene epoch, about 35 million years ago, these animals were joined by two other placental groups, the rodents and the New World monkeys. It is still uncertain whether they came from Africa, where their nearest relatives are to be found—island-hopping across the narrow ocean that then separated the two

continents—or from North America, which had provided the earliest South American mammals.

The next change to overtake South America was climatic. In the Early Tertiary the continent contained tropical, subtropical, and temperate forests and open savanna. But about 12 million years ago, in the middle of the Miocene epoch, the westward movement of the landmass brought it across the great trench in the eastern Pacific, into which the ocean floor was descending. This caused the rise of the Andes mountains along the western margin, which prevented the rain-bearing winds from reaching the rest of the continent. As it became drier, treeless pampas, cool steppe, and semi-desert replaced the savannas, leading to a reduction in the numbers of rodents, ground sloths, and hoofed mammals that had fed there.

The movement of the continent also brought it progressively closer to North America; and about six million years ago, in the Late Miocene, rodents and raccoons from North America began to appear in the south, and South American ground sloths in the north. This limited exchange suggests that there was still only an incomplete island route between the two landmasses.

A land link

When the land bridge, the Panama Isthmus, was eventually completed about three million years ago in the Pliocene epoch, there was an interchange of the two faunas. This is known as the great American interchange. At first they coexisted harmoniously throughout the tropical regions from the south of the United States to southern Brazil. But gradually the old hoofed mammals of South America either fell victim to the competent northern predators or failed to compete successfully with the immigrant herbivores. Today none of them, nor any of the giant tree sloths, exists. The opossums, armadillos, and anteaters were the only South American mammals to penetrate the colder lands of North America.

Apart from the armadillos, anteaters,

and tree sloths, South American mammals today give little hint of the unique fauna that once existed there. In contrast, the South American birds remain the richest fauna in the world, with more than 700 unique species. In particular, the vast forests of the Amazon Basin, with trees up to 330 ft tall, have provided an evolutionary hothouse for countless arboreal and flying creatures.

Until two million years ago, there was a gradual transition between the tropical faunas and floras of South and Central America and those of the cooler North American environment. But geological changes as South America continued to move westward caused the rise of the high Mexican Plateau, and this now forms a sharp boundary between the two environments. Differences in the richness of animal and plant life between the two continents were further exaggerated by the glacials that have decimated northern lands over the last two million years.

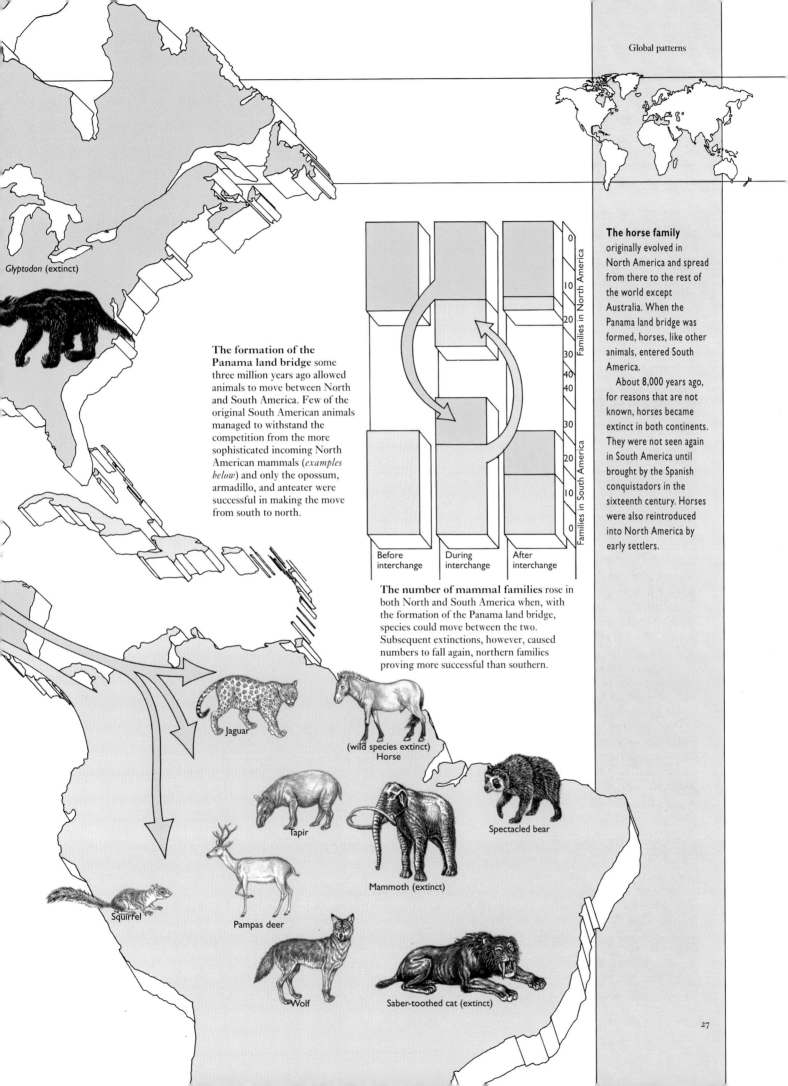

Glyptodon (extinct)

The horse family originally evolved in North America and spread from there to the rest of the world except Australia. When the Panama land bridge was formed, horses, like other animals, entered South America.

About 8,000 years ago, for reasons that are not known, horses became extinct in both continents. They were not seen again in South America until brought by the Spanish conquistadors in the sixteenth century. Horses were also reintroduced into North America by early settlers.

The formation of the Panama land bridge some three million years ago allowed animals to move between North and South America. Few of the original South American animals managed to withstand the competition from the more sophisticated incoming North American mammals (*examples below*) and only the opossum, armadillo, and anteater were successful in making the move from south to north.

Families in North America

0
10
20
30
40
40
30
20
10
0

Families in South America

Before interchange
During interchange
After interchange

The number of mammal families rose in both North and South America when, with the formation of the Panama land bridge, species could move between the two. Subsequent extinctions, however, caused numbers to fall again, northern families proving more successful than southern.

Jaguar

(wild species extinct) Horse

Tapir

Mammoth (extinct)

Spectacled bear

Pampas deer

Squirrel

Wolf

Saber-toothed cat (extinct)

27

Extinctions: the other side of evolution

Just as the death of an individual allows space for a younger, more vigorous replacement, so the extinction of a group of animals is a normal, healthy event in nature. Their disappearance is often due to competition from another group, which will replace them in the environment that they once shared. But from time to time over the history of life, the fossil record appears to show a comparatively brief period during which many types of organisms vanished. The challenge, then, is to try to discover, millions of years after the event, what may have caused these sudden waves of extinctions.

The most famous extinction event is associated with the demise of the dinosaurs at the end of the Cretaceous period, about 65 million years ago. In fact, the picture is more complicated because many other types of animals seem also to have become extinct at this time, including some families of marsupial and placental mammals, flying pterosaurs and birds, and some semi-aquatic crocodilians. There were also many extinctions in the seas and among plants.

The great variety of organisms involved, and the diversity of their habitats, suggest that the normal processes of

competition were not involved; some much more general change must have taken place. The most obvious possibility is a change in climate. Evidence from studies of floras in North America and Europe, and analyses of the plankton remains in marine sediments, suggest a substantial decline in temperatures.

The difficulty with this idea, however, is that climatic change is normally fairly slow, not sudden. But in 1981 two American scientists proposed a dramatic solution to the puzzle.

Meteorite impact

Luis and Walter Alvarez discovered that rocks laid down in many parts of the world at the extreme end of the Cretaceous contain a thin layer of iridium and osmium. Since these normally rare elements are found in larger quantities in meteorites, they suggested that a great meteor was the cause of both the iridium-osmium layer and the extinctions. They calculated that such devastation could have been caused by collision with a meteor 6 miles in diameter.

The scene they painted was awesome. The impact would have thrown up such a vast cloud of debris that the sky was darkened for years. In the prolonged and extreme "winter" that followed, plants would have failed to grow, so there was little food, and in any case background temperatures would have remained too low for cold-blooded animals to function. Could this sudden chilling, then, have been the cause of the extinctions?

The argument is a powerful one. The discovery of a crater 200 miles across in northern Yucatan, Mexico, dating from this time, has added considerable support to the proposal. The collision of this Mexican meteorite, which is indeed likely to have been 6–10 miles in diameter, would have generated a great fireball measuring hundreds of miles across, that swept over the Earth. Smoke and soot would have darkened the skies and the precipitation would have fallen as nitric acid. It is extremely probable that such a catastrophe would have caused massive extinctions over land and sea.

Whether the eventual fate of the last of the dinosaurs lay in some celestial calamity or merely in earthbound changes of climate is still unclear. What is certain, however, is that any extinction, whatever its cause, contains within itself the seeds of new life, as surviving organisms lose some competitors but themselves compete to fill the empty niches.

A complete fossilized skeleton of a long-necked dinosaur unearthed in Zigong is one of many such exciting finds in China over the last 50 years. From evidence such as this scientists can reconstruct the fauna of the different continents and their evolution.

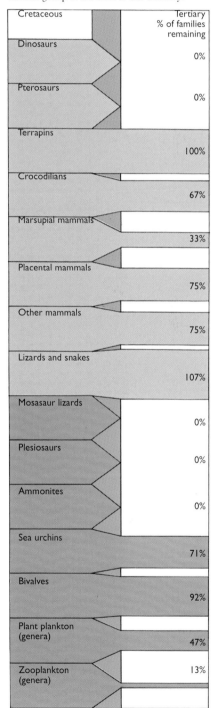

Extinctions at the end of the Cretaceous

The chart shows the percentage of families left in each group at the start of the Tertiary.

Cretaceous	Tertiary % of families remaining
Dinosaurs	0%
Pterosaurs	0%
Terrapins	100%
Crocodilians	67%
Marsupial mammals	33%
Placental mammals	75%
Other mammals	75%
Lizards and snakes	107%
Mosasaur lizards	0%
Plesiosaurs	0%
Ammonites	0%
Sea urchins	71%
Bivalves	92%
Plant plankton (genera)	47%
Zooplankton (genera)	13%

Ice ages

About 2 percent of all the water on Earth is locked up as ice, mainly in the Greenland ice cap and over the continent of Antarctica. The Antarctic ice cap is vast—one and a half times the area of the United States.

But ice caps are not a permanent feature of the Earth's geography; they have been present for only a tiny proportion of the 4.6 billion years of its existence. Four million years ago there were no ice caps in the polar regions; in fact, for most of its history the Earth has been consistently warmer than it is now. Periodically, however, it has entered into cooler states—ice ages. Ice caps have then formed and extended in area.

An "ice age" is not simply a long unbroken spell of frigid climate. Geological evidence suggests that there have been several phases during the course of the recent Ice Age when the ice caps became more extensive and glaciers overwhelmed many areas that now enjoy a temperate climate. These cold phases, known as glacials, are interspersed with warmer periods, or interglacials, such as the world is currently experiencing.

Why ice ages occur

There have been many attempts to explain the mechanism of the ice age cycle and to identify the cyclical process that reduces the Earth's energy intake from the sun so that it enters a cold period. The long-term cycle, in which ice ages are separated by hundreds of millions of years, could be accounted for by the passage of the solar system through clouds of dust in our galaxy, cutting down the amount of solar radiation reaching Earth.

This would not, however, explain the shorter-term cycle of glacials and interglacials. The best overall explanation of these cycles is still that offered by the Yugoslav physicist Milutin Milankovich in the 1930s. According to his theory there are three rhythms superimposed upon one another that together influence global climate. First, there is a cyclical variation in the orbit of the Earth around the sun. Sometimes our orbit is fairly circular, but

The ice sheets

In the last glacial period, more than half a mile of ice covered many areas that now have a temperate climate. Great ice sheets extended over northern Eurasia, North America, and Greenland.

- Extent of last great ice sheets 18,000 years ago
- Glaciation today

At least seven ice ages are thought to have occurred in the Earth's history. The last started about two million years ago and is still going on.

North Pole

PRECAMBRIAN
100 million years ago
0 1 2 3 4 5 6 7 8 9 10 ... 25 ... 45
Pleistocene, Permo-Carboniferous, Ordovician, Verangian, Sturtian, Gneisso, Huronian

Ice sheets force vegetation zones toward the Equator during a glacial. Here, strips stretching from the Mediterranean to the Arctic reveal the difference in vegetation zones during a glacial and an interglacial.

Interglacial NORTH Glacial
Arctic
Alps
SOUTH
Mediterranean

at others it becomes more elliptical, which means that at times the Earth goes farther away from the sun and becomes cooler. This elliptical cycle takes about 100,000 years to complete, roughly corresponding to the span of a glacial/interglacial cycle.

The second cycle relates to the tilt of the Earth's axis in relation to the sun, which in turn influences the seasons. This cycle, in which the tilt varies between about 22° and 24.5°, takes about 40,000 years to complete. Last, there is a cyclical wobble of the Earth's axis around its basic angle of tilt that is completed every 21,000 years. When these three cycles are superimposed, the predicted effects on climate are not unlike the actual climatic changes of the last million years.

There are, however, other factors that influence climate and that make the series

of cycles less predictable. Volcanic activity is perhaps the most important. After a severe volcanic eruption, the Earth's climate becomes cooler because of the dust veil that is forced high into the atmosphere and blocks incoming solar radiation. For example, two major volcanic eruptions in the nineteenth century, Ghaie, in the Bismarck Archipelago, and Krakatoa, caused the temperature of the northern hemisphere to fall by 2.7° Fahrenheit.

In the future human impact on the world climate may be more important than any other factor. The burning of fossil fuels is the major cause of climatic change, such as the greenhouse effect (see pp. 166–67). And a nuclear war could generate dust clouds far in excess of volcanic eruptions and lead to global cooling and a nuclear winter.

For most of the 4.6 billion years of the Earth's existence there have been no ice caps. But the Earth is currently in an ice age and major ice caps are present in Greenland and Antarctica.

Survivors of the Ice Age

The climatic changes of glacial episodes, the coldest phases of an ice age, affected more than just the high latitudes to the far north and south. The climate of the whole Earth was changed, and consequently the pattern of vegetation over the surface of the globe was altered.

As the coniferous and deciduous forests of the temperate zones were pushed toward the equator by the advancing ice, an area of dry grassland and scrub preceded them. The equatorial regions, occupied today by tropical rain forest, were somewhat drier during the glacials, and the forest itself became fragmented into smaller units. These were interspersed with tropical savanna grasslands. Trees from mountain regions became more abundant in the lowlands.

The changing patterns of climate and vegetation presented many problems for the various species of plants and animals that had to cope with these fluctuations. Some became extinct; others became much more restricted in their distributions. As the climate cooled, Northern Hemisphere plants were able to germinate and establish themselves south of their usual ranges. But northerly populations were gradually exterminated.

In Europe the stresses experienced during the glacial advances were, in some respects, more severe than in North America. The mountain ranges in Europe, the Alps and Pyrenees, run approximately east to west and are high enough to have developed their own ice masses, cutting off the southward spread of the plants and animals threatened by the advancing ice of the north. Some trees, such as the hemlock and the tulip tree, became extinct in Europe during successive glacials but survived in North America where their freedom to spread their range was not restricted by mountain ranges.

Gorillas

The gorilla (*Gorilla gorilla*) lives in the African tropical rain forest. Fragmentation of the rain forests during the climatic changes of the Ice Age is probably the main cause of the split in their distribution into separate western and eastern populations. The gorillas took refuge in the surviving patches.

Although, after the glacial, the forest recovered, the appearance and gradual development of the Zaire River and the more recent habitat destruction by humans in the intervening area have prevented the two populations from fusing once more. Although still the same species, they have evolved into two distinct races and will eventually, if they survive, become separate species of gorilla.

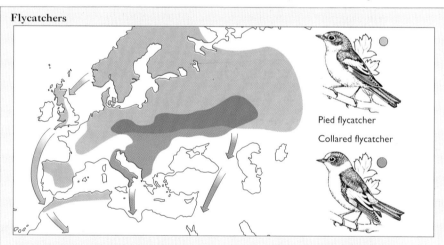

Flycatchers

Pied flycatcher

Collared flycatcher

The European and Asian flycatchers are small migratory birds that feed on insects. They move from Africa, where they spend the winter, into Europe and Asia to take advantage of the longer days and rich feeding opportunities afforded by the temperate summer.

Both the pied flycatcher (*Ficedula hypoleuca*) and the collared flycatcher (*Ficedula albicollis*) occur in Europe in summer, but the latter has a more southerly distribution. The two do not interbreed, despite their similarity, but they are clearly closely related.

In the relatively recent past, less than one million years ago, these flycatchers were a single species. Their migration patterns may provide a clue to their separate evolution. Pied flycatchers head west in the fall and cross the Mediterranean into Africa by moving through southern Spain. Collared flycatchers, however, migrate east through Turkey and the Middle East. These habits were probably developed during the last Ice Age when their breeding ranges were separated by the ice sheet and the two flycatcher populations did not meet at all during the breeding seasons.

The flycatchers evolved independently, and now, even though the two species are back in contact again, they show little interest in each other and retain their old migratory habits. So glaciation proved a spur to evolution, with the result that one species has now become two.

This disruption and fragmentation of the range of a plant or animal may result in new bursts of evolution as well as extinctions. If a species is split into separate ranges for long enough it will eventually evolve into two separate species. New species were probably born during the Ice Age; others began to split and are on their way to becoming new species.

Fast-breeding species, such as birds and pupfish of the genus *Cyprinodon* (below), evolve quickly. Large animals, and trees, which may take several decades to reach breeding maturity, may take thousands of years to evolve new species.

Past and present extent of rain forest in South America.

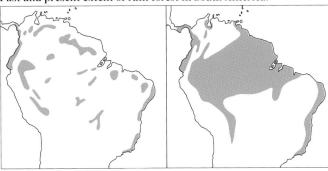

During the last glacial, rain forests were far less extensive than they are today. Climatic change caused them to become fragmented into scattered "refuges," where conditions remained sufficiently warm and moist.

Desert pupfish

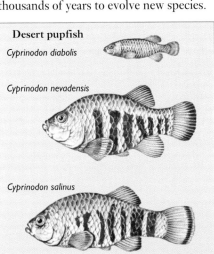

Cyprinodon diabolis

Cyprinodon nevadensis

Cyprinodon salinus

In the Nevada desert small lakes and waterholes provide homes for about 20 species of desert pupfish (*Cyprinodon* spp.). Males are iridescent blue, females are green, and there is a strong resemblance between species.

At the end of the last glacial, conditions in the area, and in most of the currently arid regions of the world, were much wetter. Connections between the various bodies of water allowed fish to move around and interbreed. As the glacial passed and the climate became more arid over the last six or seven thousand years, lakes became smaller and more isolated. The pupfish split into scattered populations, and each became a separate species.

White spruce

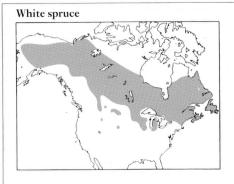

The white spruce (*Picea glauca*) of North America mainly occurs in a broad band across Canada and into Alaska. But there are isolated populations of the tree in hilly areas south of its main range, notably in the Black Hills of Dakota.

This apparently puzzling distribution can best be explained by the tree's history. Having been driven south during the last glacial, these Dakota populations survived as relicts and are now isolated from the main band of distribution. This can be confirmed by the fossil record. The spruce produces large quantities of distinctively shaped pollen grains that have been preserved in lake sediments and provide evidence of the tree's former southerly distribution.

The Dakota spruce has not yet evolved into a new species. Its breeding rate is slow and the population has been isolated for a relatively short time.

During an ice age the climate changes worldwide. In the last glacial the American Midwest was some 18°F colder than it is today, and Britain about 12.5°F colder.

The Atlantic was affected more severely than the Pacific. The sea around Ice Age Spain was about the same temperature as the waters surrounding modern Greenland.

Unlike animals, plants could not move toward the equator as the Earth grew cooler, and many populations of trees perished as the ice advanced.

But if the seeds of these trees had dispersed southward, they were able to germinate where once it had been too hot and dry; in this way plants could spread south and escape the advancing ice.

Trees can migrate at rates of several hundred feet a year. The hazel tree achieved rates of 5,000 ft per year when spreading north after the last glaciation.

Ice Age devastation

At the end of the last glacial, about 10,000 years ago, more than twice as many species of animals and plants became extinct as at any previous period of glacial advance and retreat. The most spectacular extinctions involved the so-called "megafauna," animals whose body weights were in excess of 110 lb. Most were mammals, among them mammoths, mastodons, cave bears, ground sloths, saber-toothed cats, and giant kangaroos, but there were also some birds. It has been estimated that at least 55 species of large animals became extinct at this time in North America alone.

What could have caused such an extraordinary loss of species? Was it simply the consequence of an unstable climate causing changes in habitats, or were other factors at work? Specifically, to what extent did people bring about these extinctions?

If human predation was indeed implicated, the disappearance of an animal could be expected to correspond with the arrival of people. But determining such dates is difficult. For example, it can never be conclusively asserted that the last bones of a species have in fact been found.

A better source of evidence is often the animal's dung—a big herbivore produces much more dung than bones in the course of its life. Preservation may be a problem, but dung survives well in caves, especially in arid areas. It has proved useful, for example, in dating the demise of the large, lumbering Shasta ground sloth in the southwest United States. For many sites examined, the last traces can be dated to about 11,000 years ago—a figure that corresponds closely with the arrival of hunting cultures.

Climatic change or human intervention?

The spread of scrub and forests undoubtedly contributed to the extinction of species adapted to the tundra, such as the woolly rhinoceros, woolly mammoth, cave bear, and giant Irish elk. But is it possible that climate and habitat changes simply rendered these animals more sensitive to further pressures, such as hunting?

There is, of course, plentiful direct evidence of human predation on the megafauna. In Asia, the mammoth was not only a source of meat but also of bones for the construction of dwellings. In the caves of southern France and of Spain, the hunters themselves left a vivid pictorial record of their prey and their methods. Bison, horses, wild cattle, and reindeer are all represented graphically on the walls.

This does not necessarily mean that the apparently unrelated extinction of smaller species at this time is to be explained in terms of climatic change rather than human intervention. Changes in habitats, induced by humans, could account for these losses too. Some transformations could have been direct—as in the case of the destruction of forested land by fire. An

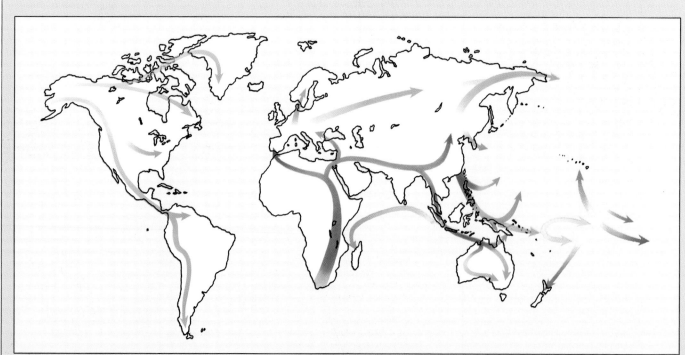

Our species is believed to have arisen in Africa and spread through much of the rest of the world during the last glacial, ending ten thousand years ago. The arrival of *Homo sapiens* often seems to have coincided with the extinction of large animals.

example of an indirect change might be the invasion of scrub on grassland, thanks to the hunting to extinction of a "keystone species"—a particularly influential animal such as a large grazer. Just such an effect can be seen in Africa today wherever the elephant has disappeared from the savanna.

Additional evidence that points emphatically to humans as the culprits in the extinction story derives from places where their arrival is recent and can be dated with some confidence. Examples include New Zealand, where 27 species of flightless birds, including a giant moa standing 10 ft high, became extinct soon after the appearance of people there about 1,000 years ago.

While other factors, such as climatic change, habitat alterations, and even disease, may have contributed to megafaunal extinctions, there is strong circumstantial evidence, though no conclusive proof, of human responsibility.

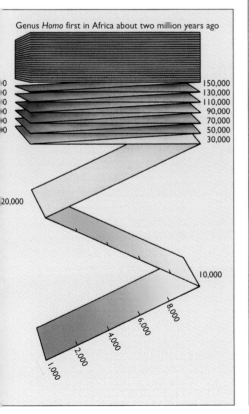

Genus *Homo* first in Africa about two million years ago

150,000
130,000
110,000
90,000
70,000
50,000
30,000
20,000
10,000
8,000
6,000
4,000
2,000
1,000

Many large animals disappeared from the face of the Earth at the beginning of the present warm period. They had survived previous warm periods, but this one was different for an important new predator was present—humankind. Because these large creatures moved slowly, they easily fell victim to human hunters. (Color codes *below* link with time chart and map, *left*.)

26,000–15,000 years ago
○ AUSTRALIA
Diprotodon
Giant kangaroos

12,000–9,000 years ago
○ SOUTHERN AFRICA
Megalotragus
Metridiochoerus
Equus capensis
Antidorcas australis
Antidorcas bondi

11,000 years ago
○ EUROPE/ASIA
Mammuthus
Coleodonta
Elasmotherium
Megaloceros
Ovibes

11,000–10,000 years ago
○ NORTH AMERICA
Mammuthus
Mammut americanum
Cuvieronius
Eremotherium

11,000–10,000 years ago
○ SOUTH AMERICA
Cuvieronius
Haplomastodon
Stegomastodon
Eremotherium
Megatherium
Toxodon

8,000 years ago
○ SICILY
Elephas falconeri

2,000–1,000 years ago
○ MADAGASCAR
Dwarf hippopotamus
Giant lemurs
Elephant bird (Aepyornis)

1,000 years ago
○ NEW ZEALAND
Moa and other flightless birds

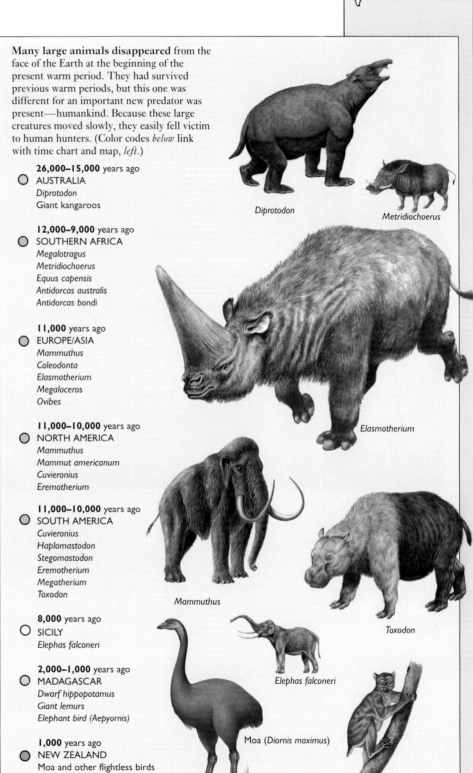

Diprotodon

Metridiochoerus

Elasmotherium

Mammuthus

Toxodon

Elephas falconeri

Moa (*Diornis maximus*)

Giant lemur (*Megaladapis*)

Birth of
an island chain

The Hawaiian Islands, clothed in dense green vegetation and surrounded by white breakers, rise from the blue waters of the Pacific Ocean. Only the slow drift of smoke from their few, snowcapped volcanoes hints at the dramatic story of their birth in fire, molten lava, and hissing steam.

The unique chain includes not only the eight major islands from Hawaii itself westward to Nihau, but the smaller Leeward Islands. These smaller islands, islets, and reefs stretch even farther west to Midway Island and Kure, 1,400 miles from Hawaii. Extending northwestward from Midway almost to the Aleutians is the Emperor Seamount chain of 30 submerged seamounts. Were the oceans miraculously to dry up, the Hawaiian Islands would be seen as the most impressive mountains on Earth; they rise to

nearly 6 miles from the ocean floor.

All the islands of the Hawaiian chain are volcanic and can be dated by scientists from the radioactive minerals that they contain. Hawaii, the largest and most easterly island, is the youngest; its surface rocks are less than a million years old. From there, the farther west the island, the greater is its age. Oahu is 2.6 to 3.6 million years old, Kauai about 5 million years old, and the most westerly island, Midway, is 27 million years old.

The pattern continues through the northwesterly-directed Emperor Seamount chain: the farthest north, the Meiji Seamount, is the oldest of them all at 70 million years. But the ocean floor from which the chain rises is in fact far older, 80 to 120 million years, which poses the question of how this neat pattern of islands has been formed.

Hotspots of volcanic activity

Geologists have discovered more than 120 "hotspots"—sites of intense volcanic activity—scattered over the surface of the globe. More than 50 of these lie below the oceans; the remainder are below continents in areas such as East Africa and Iceland. A hotspot is probably the cause of volcanic activity in Yellowstone Park in the United States, while another has given birth to all the islands, islets, and seamounts of the Hawaiian-Emperor series.

The unusual feature of the Hawaiian hotspot is that, because of the pattern of plate tectonics (see p. 14), the floor of the Pacific Ocean is steadily moving over it at a speed of some 3–4 in a year. The hotspot is probably about 185 miles in diameter and lies at least 35 miles below the ocean floor. The floor over the hotspot becomes heated and swollen and eventually a rift

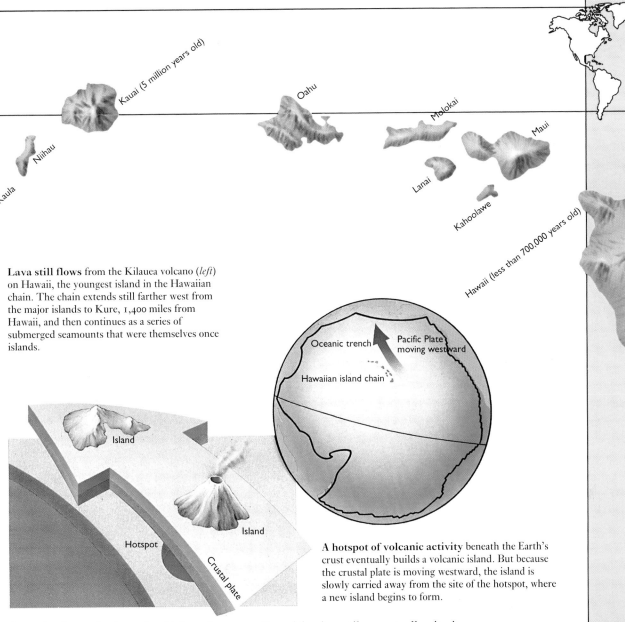

Kauai (5 million years old)

Oahu

Molokai

Maui

Niihau

Lanai

Kaula

Kahoolawe

Hawaii (less than 700,000 years old)

Oceanic trench

Pacific Plate moving westward

Hawaiian island chain

Island

Hotspot

Island

Crustal plate

Lava still flows from the Kilauea volcano (*left*) on Hawaii, the youngest island in the Hawaiian chain. The chain extends still farther west from the major islands to Kure, 1,400 miles from Hawaii, and then continues as a series of submerged seamounts that were themselves once islands.

A hotspot of volcanic activity beneath the Earth's crust eventually builds a volcanic island. But because the crustal plate is moving westward, the island is slowly carried away from the site of the hotspot, where a new island begins to form.

forms in the rocks through which molten lava rises. At first this creates only a hot vent on the ocean floor. But as the years pass, the submarine volcano increases in size until it eventually breaks through the surface of the sea to form a volcanic island.

The new island continues to grow as long as it is connected to the hotspot. The island of Hawaii is still at this stage and its southeastern volcanoes, Kilauea and Mauna Loa, are still producing lava.

Eventually, however, the slow movement of the Pacific Plate westward will separate the new volcanic island from its source of new lava. Then the relentless forces of erosion by rain, wind, and ocean waves, which have always been at work, will steadily wear away the island, reducing its height and size. In addition, as the ocean floor moves away from the hotspot, it will cool and contract, pulling down the

eroding island until eventually it is reduced to a mere sunken seamount, deep below the surface of the sea.

The seamount relics of ancient volcanic islands are still being borne along on the moving Pacific Plate until they reach one of the great oceanic trenches into which the old ocean floor disappears down into the deeper layers of the Earth to be recycled. In a few million years the oldest seamount, Meiji, will reach the edge of the great trench south of the Aleutian Islands and disappear forever.

The whole cycle repeats endlessly. Once Hawaii has finally passed from the hotspot, it will commence its 70 million-year-long journey westward toward eventual destruction and oblivion. And a bulge 1,000 ft high in the ocean floor southeast of Hawaii will one day be the site of a new island.

Hawaiian colonists

The Hawaiian Islands are the most isolated in the world. The largest, Hawaii itself, is 2,000 miles from North America and 3,400 miles from Asia.

Some organisms have crossed the vast stretches of ocean to these remote tropical islands. And much can be learned from the evolutionary history of those that succeeded—the animals and plants that established themselves in a new world of opportunity, far from their normal competitors, predators, and parasites.

Few of the Hawaiian chain's land animals and plants come from the distant North American or Asian mainlands. Most are related to forms on other Pacific islands, which are geographically closer to Hawaii. Moreover, the presence of those animals and plants on other islands shows that they were already well adapted to island life and to overseas dispersal.

An opportunity for evolution

Many types of plant and animal never reached the Hawaiian chain. There are no naturally occurring freshwater fishes, amphibians, reptiles, or land mammals, and no coniferous trees. And of those groups that did reach the islands, representatives of only a few families finally succeeded in their colonization. The failure of the majority, however, provided the successful colonists with an unparalleled opportunity to evolve and diversify. The limited range of birds and insects, for example, filled all the ecological niches that their relatives had occupied in other lands. Such diversification is known as adaptive radiation.

One of the species of birds that reached the islands, probably about 15 to 20 million years ago, was finchlike. Its descendants form a unique family, the honeycreepers (Drepanididae), of which there are now 23 species. They have evolved to fill a variety of ways of life, feeding on insects, insect larvae, fruits, seeds, and nectar.

Among plants, the lobelias (Campanulaceae) have evolved in a variety of habitats. About four types of these normally small herbaceous flowering plants probably colonized the Hawaiian Islands originally. These have radiated into six genera, including over 150 species and varieties.

Perhaps the most impressive examples of adaptive radiation on the Hawaiian Islands are the tiny fruit flies of the family Drosophilidae. More than 500 species of these flies live in the islands, nearly a quarter of the total known throughout the world. This diversity is probably due partly to the great variation in climate and vegetation. In addition, lava-flows frequently isolate small patches of vegetation—mini-islands—in which evolution then takes place independently.

Genetic studies of the fruit flies have shown that the relationships between them reflect the geological history of the islands. For example, species from the older islands, to the west, are found to be ancestors of those in the younger islands to the east.

As in other parts of the world, the animals and plants native to the islands are increasingly threatened by the activities of humans, and by the animals and plants that they import. Human disturbance and the effects of introduced species have led to the loss of many of the islands' species of birds. Efforts are being made to protect and conserve this unique living museum, but for some of its inhabitants this concern has certainly come too late.

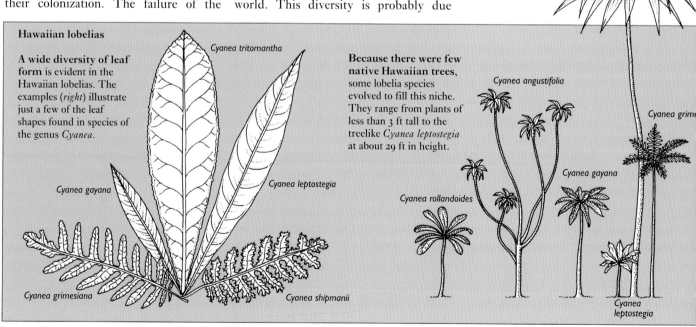

Hawaiian lobelias

A wide diversity of leaf form is evident in the Hawaiian lobelias. The examples (*right*) illustrate just a few of the leaf shapes found in species of the genus *Cyanea*.

Because there were few native Hawaiian trees, some lobelia species evolved to fill this niche. They range from plants of less than 3 ft tall to the treelike *Cyanea leptostegia* at about 29 ft in height.

Cyanea tritomantha

Cyanea gayana

Cyanea leptostegia

Cyanea grimesiana

Cyanea shipmanii

Cyanea angustifolia

Cyanea grime

Cyanea gayana

Cyanea rollandoides

Cyanea leptostegia

waiian honeycreepers

Fruit | Fruit and seeds | Insects | Insects and some nectar | Nectar and some insects

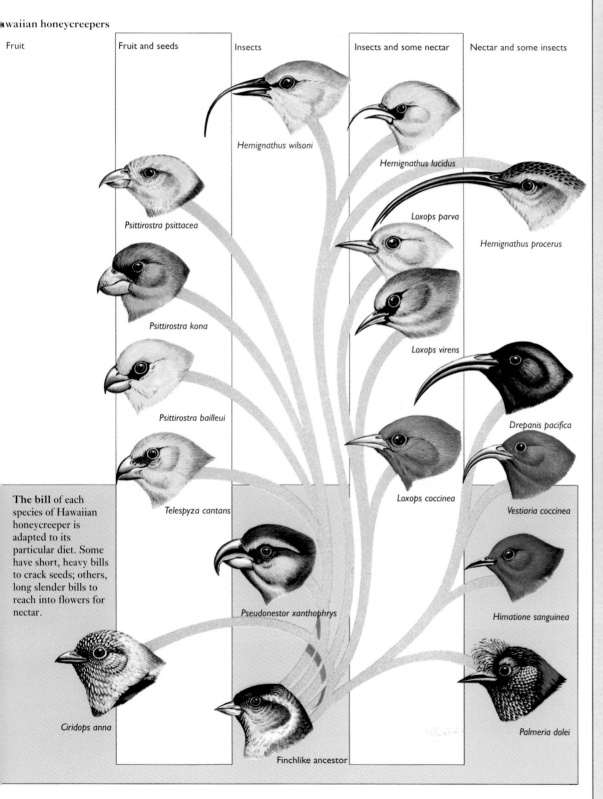

Hemignathus wilsoni

Hemignathus lucidus

Psittirostra psittacea

Loxops parva

Hemignathus procerus

Psittirostra kona

Psittirostra bailleui

Loxops virens

Telespyza cantans

Drepanis pacifica

The bill of each species of Hawaiian honeycreeper is adapted to its particular diet. Some have short, heavy bills to crack seeds; others, long slender bills to reach into flowers for nectar.

Loxops coccinea

Vestiaria coccinea

Pseudonestor xanthophrys

Himatione sanguinea

Ciridops anna

Finchlike ancestor

Palmeria dolei

39

Worldwide patterns

Each species of animal and plant has its own distinctive pattern of distribution over the face of the Earth. No two patterns are exactly the same. Some species or groups of species enjoy an almost complete global coverage, whereas others are restricted, sometimes to a very small area. Others, perhaps the most interesting of all, have split distributions, sometimes occurring in widely separated parts of the world.

Explaining a pattern of distribution for an organism involves knowing the answers to many different questions. How tolerant is it of different physical factors in the environment? How adaptable is it to new conditions? When did it evolve and what was the Earth like at that time? How well can it compete with other organisms? What was its past distribution, and how did it fare during the climatic changes of the recent past such as the last glacial? All of these questions are relevant to an understanding of the modern geography of plants and animals.

The adaptable plantain
The widespread, or cosmopolitan, type of organism is usually tolerant of a wide range of environmental conditions. The broad-leaved plantain (*Plantago major*), for example, is found over most of the globe and is tolerant of heat and cold, wetness and drought, and, perhaps most important of all, disturbance by mankind. Indeed, it benefits from human company because the habitat disturbance caused by humans reduces the amount of competition from other, especially shade-casting, plant species. Even the cosmopolitan species have their weaknesses and plantains are not tolerant of shade.

Among animals, a wading bird, the common snipe (*Gallinago gallinago*), is found in every continent except Antarctica and Australasia. It has a long bill and legs, enabling it to feed on small invertebrates in shallow water, and its tastes are catholic. It consumes a very wide range of invertebrates from worms to dragonfly larvae and has been known to eat the seeds of wetland plants. This wide choice of food certainly contributes to its success

around the world. An animal that is not fastidious in its diet is more likely to be able to occupy a range of habitats.

Both the snipe and the plantain, therefore, are extremely tolerant and adaptable in their physical and food requirements. Both also have efficient means of dispersal. The plantain has seeds that are consumed by birds and germinate the better for having passed through the bird's digestive system. The snipe can fly. Many of the most successful cosmopolitan plants have developed close relationships with highly mobile animals, especially migratory birds.

Cosmopolitan families
Individual species with cosmopolitan distributions are, however, comparatively rare. But at a higher level of classification, while an individual species may not be found throughout the globe, the family or order to which it belongs does have a worldwide distribution. The crow family (Corvidae) is represented on all continents but no single species is cosmopolitan. Among reptiles, the skinks (Scincidae) are widespread in all but the high latitudes.

Several plant families are cosmopolitan, and probably the most widespread is the grass family (Gramineae). This is the flowering plant family that extends closest to the South Pole.

When looking at the distribution of families rather than species, the length of time for which that family has been present on Earth must be a factor. A family represented on all major continents could have evolved and been represented in each region before the great splitting up of continental masses. This may well have occurred with the rats and mice, and the roses, both ancient families. However, both have good powers of dispersal so they may have continued to spread.

With such mobile creatures as skinks and crows, dispersal over a lengthy period is entirely possible and geological explanations of their widespread distribution are hardly necessary. The same applies to our own species, *Homo sapiens*, with one of the most cosmopolitan distributions of all.

Plantain

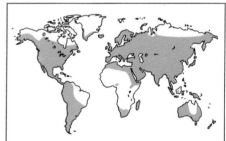

The broad-leaved plantain (*Plantago major*) is a successful weed that occurs in every continent. Like most widespread plants, it is adaptable. Plantain will tolerate most conditions other than shade. It thrives in habitats disturbed by humans, where many other plants have been destroyed.

The grasses of the family Poaceae are one of the most successful of all flowering plant groups. Representatives of the family are found in almost every habitat, from alpine pastures to coastal mudflats, from tropical savannas to Antarctic islands.

As much as 20 percent of the Earth's land surface is dominated by grasses. These plants can cope with all types of environmental stress; they can also survive being grazed on by animals because they grow from the base rather than the tip.

Ferns are among the most efficient of plants at dispersing themselves. They germinate from minute spores carried in the air as dust.

The world's most widespread fern is probably the brittle bladder fern (*Cystopteris fragilis*), found from Greenland to tropical mountains.

No barrier is too high or too wide for the spores of this plant to cross. It was among the first land plants to colonize the volcanic island of Surtsey when it arose off the coast of Iceland in 1963.

Snipe

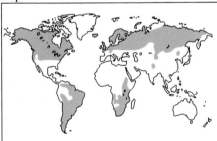

Skink family

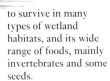

The snipe (*Gallinago gallinago*) is a wetland bird found in all but the most arid areas of the world. Its success is due both to its ability to survive in many types of wetland habitats, and its wide range of foods, mainly invertebrates and some seeds.

The skink family (Scincidae) is a particularly successful and widespread group of reptiles. They occur in every continent except Antarctica, and thrive in a wide variety of habitats, from rocky shores to desert, and from freshwater to forest.

Restricted patterns 1

Some plants and animals are widespread throughout the world, but most are restricted in their distribution, and the endemics (see pp. 38–39) are usually found only where they originated. The reasons for their distribution patterns lie in the geography, climate, or competition existing when the organisms evolved, which may have been at any time in the last 100 million years.

Drifting continents

Changing geography has been one of the most important causes of discontinuous, or disjunct, distribution. When the great continental masses broke up and slowly rafted apart, each fragment still bore its original flora and fauna. This is illustrated in the breakup of Gondwanaland, the southern supercontinent, which occurred during the Cretaceous period.

The flightless birds belonging to the order Ratites are today distributed across many continents: the rhea in South America, the ostrich in Africa, the cassowary and emu in Australia, and the kiwi in New Zealand. Since it is impossible for these birds to have crossed the oceans separating the continents, they must have originated in Gondwanaland and only later have evolved into different genera.

Not all the southern-hemisphere families developed before the breakup of Gondwanaland was complete. Both the marsupial mammals and the southern beech tree, *Nothofagus*, evolved after Africa and India had separated off, so they are found only in South America and Australia and in fossil form in Antarctica.

New land connections also subdivided marine fauna. When the Panama Isthmus finally connected North and South America in the Pliocene period (see pp. 26–27), it separated a single marine fauna into Pacific and Caribbean parts. Hundreds of species of marine organisms segregated in this way, including crabs, mollusks, and fishes, have since evolved into pairs of species, one on each side of the isthmus.

Another cause of disjunct distribution patterns is competition: a group originally widespread across the world may have died out because it could not compete with later-evolving organisms. Fossil evidence indicates that the primitive cycad plants, for instance, once grew worldwide; now they are found only in Central America, northern South America, Africa, Madagascar, and Australasia. They were presumably killed off in other places by

The spectacular bloom of an Australian species of the Protea family.

competition from flowering plants.

The progressive cooling of the climate that took place during the Cenozoic period also had an effect on the distribution of organisms, particularly in the northern hemisphere. Forests containing a diversity of coniferous and broad-leaved flowering trees had evolved in all the continents there and were once found as far north as Alaska. As the climate cooled, however, these forests gradually shifted southward, although they remained in a wide belt across North America and Eurasia. But when movements of the tectonic plates raised mountains in western North America, southern Europe, and the Himalayan region, some trees became extinct over wide areas. Hence the irregular distribution today of the tulip tree (*Liriodendron*), and the magnolia, which are now found naturally only in eastern Asia and eastern North America.

A similar combination of climate and land movement was probably responsible also for the present disjunct distribution of tapirs (piglike mammals). These animals were once found throughout North America and tropical Eurasia. But the cooling of the northern hemisphere led to the extinction of European tapirs, which were unable to cross the Mediterranean Sea or the Sahara Desert to reach the tropics of central Africa.

Their cousins in North America were more fortunate, for the completion of the Panama Isthmus allowed them to colonize South America before they became extinct in the north. So, today tapirs are found only in South America and Southeast Asia.

Iguanid lizards and bold snakes are found today in both the New World and Madagascar but not in intervening Africa.

A possible explanation for this is that both groups did once exist in Africa but have become extinct there. This was probably due to competition from their relatives, the agamid lizards and pythonid snakes, which are found in Africa but not in the Americas or in Madagascar.

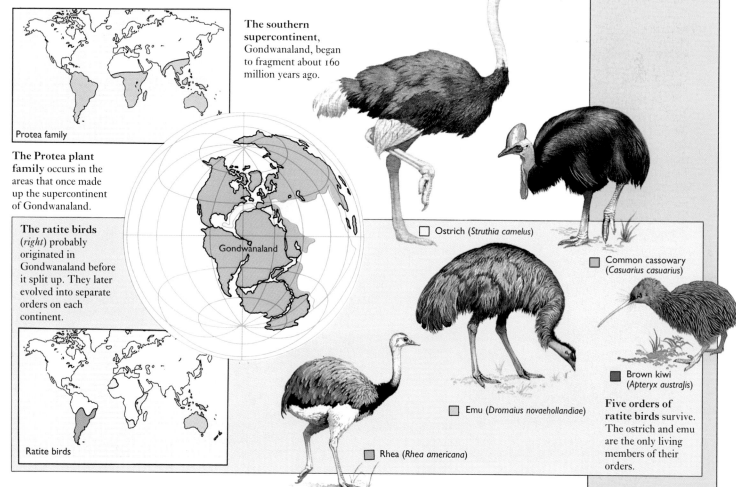

Protea family

The Protea plant family occurs in the areas that once made up the supercontinent of Gondwanaland.

The ratite birds (*right*) probably originated in Gondwanaland before it split up. They later evolved into separate orders on each continent.

Ratite birds

The southern supercontinent, Gondwanaland, began to fragment about 160 million years ago.

Gondwanaland

☐ Ostrich (*Struthia camelus*)

☐ Common cassowary (*Casuarius casuarius*)

☐ Emu (*Dromaius novaehollandiae*)

■ Brown kiwi (*Apteryx australis*)

☐ Rhea (*Rhea americana*)

Five orders of ratite birds survive. The ostrich and emu are the only living members of their orders.

Restricted
patterns 2

While some plant and animal species are restricted in their range because of ancient geological events, such as the movements of continental landmasses, the spread of others has been limited by different, more recent influences. Of these, sensitivity to climate is one of the most common.

All species have their limits. The mosquito *Aedes aegypti*, for example, is restricted to tropical and warm-temperate parts of the world: neither adults nor larvae can tolerate cold. Within this range, however, the insect is extremely widespread. At another extreme, the brook saxifrage (*Saxifraga rivularis*) is restricted to arctic and subarctic regions and high mountains farther south. This plant is unable to survive high temperatures.

Even good powers of dispersal do not guarantee wide distribution if a species has climatic limits. Many tropical plants rely on marine currents to disperse their seed-containing fruits, which are often large and robustly built to withstand long immersion in seawater. Fruits of such tropical species as the sea beans frequently drift from the Caribbean across the Atlantic to western Ireland, but they are unable to germinate and establish themselves in its cooler climate.

Similarly, many birds manage to cross the Atlantic from west to east as a result of the prevailing wind directions. North American species, from waders such as dowitchers to passerines such as the yellow-billed cuckoo and the song sparrow, make the journey to Europe but do not succeed in establishing themselves there. Their failure probably has less to do with climate than with the problem of finding a mate and surviving the competitive pressures of native species.

Some species are well able to spread within a given area but are unable to cope with a particular barrier, such as a mountain range or desert. The Arabian bustard (*Ardeotis arabs*), for example, inhabits dry scrub and open savanna woodland in a belt across Africa south of the Sahara, and through Ethiopia and Somalia into southeastern Saudi Arabia. Its range is un-

doubtedly limited by the barrier of the deserts of northern Africa and Arabia.

Oceans may prove an insuperable barrier to plants. Members of the cactus family (Cactaceae), for example, are capable of living outside their current New World range, but none has ever succeeded in crossing the Atlantic or Pacific Ocean without human help.

Changing global climate, especially over

the past two million years, has led to the constriction of some species ranges. The coast redwood (*Sequoia sempervirens*) of California had an extensive distribution in western and central North America as recently as Tertiary times, two million years ago. Its natural population is now restricted to a strip of land just 25 miles wide along the fog belt of the Pacific coast.

Sometimes climatic change has even

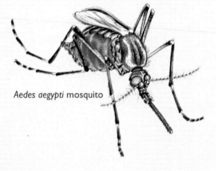

Aedes aegypti mosquito

Aedes aegypti mosquito

Colville barrel cactus

The mosquito *Aedes aegypti* occurs mainly within the area where July temperatures average at least 50°F. Larvae and adults die at temperatures below this. The species does succeed in living farther north in some regions, such as eastern North America, where it survives the winter in egg form.

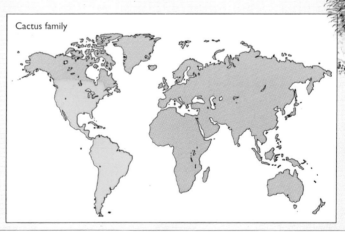

Cactus family

The cactus family occurs naturally only in the New World. Cacti are capable of living in the Old World but are thought to have evolved at a time when the Atlantic Ocean was already too wide for them to cross.

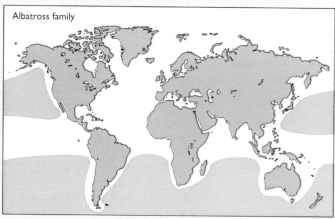

Albatross family

Albatrosses spend much of their lives on the wing but their flight is very dependent on wind. They have great difficulty in crossing the equatorial regions that experience long periods of calm (the doldrums). As a result most species live in the southern hemisphere, although three species occur in the North Pacific.

caused populations of a species to fragment. The brush mouse (*Peromyscus boylei*), for example, has its main range in New Mexico and areas to the south but is also found in a part of Texas and in several small pockets between the two areas. The distribution was once continuous but has been fragmented by climatic and habitat changes.

One further reason for a species being restricted in its distribution is that it has only recently evolved and simply has not had time to spread. California is notable for its richness in endemic species—more than 2,000 species of its flowering plants occur only in this region. Many of them, such as species of *Clarkia*, *Mimulus*, and *Cryptantha*, are newly evolved and may be awaiting an opportunity to spread beyond their current boundaries.

45

From equator
to pole

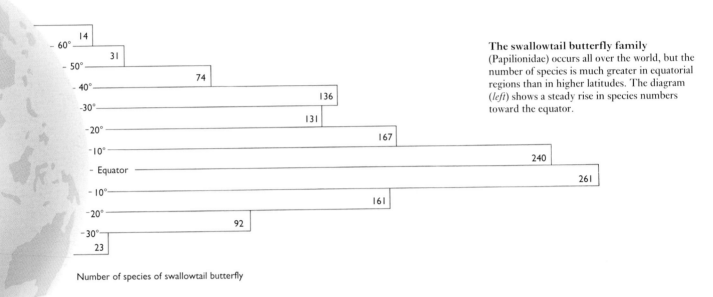

14
— 60°
31
— 50°
74
— 40°
136
-30°
131
- 20°
167
- 10°
240
- Equator
261
- 10°
161
- 20°
92
-30°
23

Number of species of swallowtail butterfly

The swallowtail butterfly family
(Papilionidae) occurs all over the world, but the number of species is much greater in equatorial regions than in higher latitudes. The diagram (*left*) shows a steady rise in species numbers toward the equator.

The luxuriant jungles of the equatorial regions of the world teem with life, in sharp contrast to the silent, lifeless icy wastes of the poles. As a rule, the nearer the equator, the greater is the diversity of species, and this seems generally to apply to all forms of life from algae and lichens to flowering plants, from insects to birds and mammals. That this gradient of diversity exists may seem obvious, but the reason why it exists is complicated and has proved hard to explain.

One approach might be to try to identify the ways in which the richest of all the biomes, the jungle or tropical rain forest, differs from the others. An astonishing feature is the immense variety of its plant life—the Amazon rain forest, for example, contains more than 4,000 species of trees. Here, an area of less than 2 acres may contain 400 trees, belonging to almost 90 species. These trees provide support for other plants, each having space for up to 80 species of epiphytic plant on its trunk and branches.

At one time it was thought that the rain forest was a community of great antiquity and stability, unaffected by the ice ages that decimated higher latitudes, but it is now known that these tropical regions did not go unscathed. Studies suggest that while higher latitudes had cycles of glacial and interglacial periods, the tropics had similar cycles of wet and dry periods. During the latter, the great forests contracted to a number of smaller patches, separated by dry shrublands (p.33). This may, indeed, be one explanation for the diversity of plants: evolution would have continued in each of these patches of forest. When wet, humid conditions returned, the new plants may have dispersed outward from the "patches" to colonize surrounding areas.

A multilayered world

Another remarkable feature of the rain forest is that it has many layers. The main tree canopy is 80–100 ft above ground, while individual trees rise to heights of half as much again, or even twice as high. Below the main canopy, smaller trees and shrubs provide another layer of life. Each layer, the forest floor included, offers a distinctive environment with a unique community of animals and plants.

The diversity of plant life provides a huge range of foods for animals. And, in the unvarying, seasonless climate, fruits and seeds are always available. Thus the animals that feed on them can become more specialized, allowing a greater range of animals in each way of life.

With its complex structure and year-round production of leaves, fruits, and seeds, the rain forest clearly provides an opportunity for the evolution of many more types of animals than do other biomes. But what is it that allows a greater diversity of plants to survive in the tropics than in higher latitudes?

The answer may be simple—sunshine. The tropics receive a constant high amount of solar energy. This allows plants to grow more rapidly and to greater heights. And since the tropics are far wealthier in solar energy than the higher latitudes, they can afford to spread that wealth over a greater diversity of species.

Not every tropical environment is as diverse as the Amazon rain forest. Sunlight alone is not enough. If rainfall or the supply of nutrients is inadequate, then diversity will remain low, as in parts of Australia.

Seasonal stresses

In higher latitudes, the supply of solar energy is not only lower, it is also irregular. In some seasons the stress of short day length and low temperatures prohibits the survival of organisms that are not hardy. If the Earth's axis were not at an angle, if it lay vertical to its orbit, the high latitudes would be seasonless and might well exhibit a higher diversity of species.

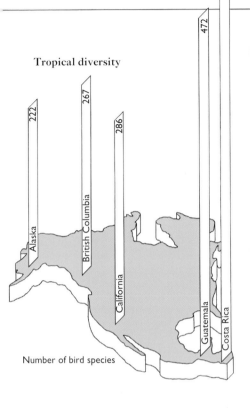

Tropical diversity

603
472
267
222
286

Alaska
British Columbia
California
Guatemala
Costa Rica

Number of bird species

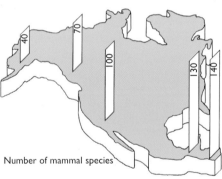

40
70
100
130
140

Number of mammal species

The number of both bird and mammal species also increases dramatically between high latitudes and the tropics in North and Central America. There are three times as many species in the tropical forests of Panama as in Alaska. Much of the increase in mammal diversity is accounted for by the numbers of bat species, which, like birds, can easily fly from tree to tree. Many feed on the abundant insects that live on the year-round diversity of tropical fruits, flowers, and foliage in the forest.

The extraordinary diversity of plant life in the tropical rain forest provides a huge range of foods and habitats for animals such as this red-eyed leaf frog in the Costa Rican forest.

Habitat
patterns

The realms of nature

Many of us today know first-hand how varied our plane We can step from the chill of northern Europe into a airplane and, within a few hours, be basking in the Mediterranean sunshine. Closer to home, but still within a we can fly from the snows of the Rocky Mountains to the baking heat of Mexico.

These great variations in the Earth's environments stem just one simple circumstance, the way the light from the su strikes the surface of the planet. On the equator, it does so square. There, at midday, the shadow of a palm tree forms black disk on the ground centered around its bole; in the de rocks get so hot that they are painful to the touch; and wate from tepid rivers vaporizes and saturates the air. On the otl hand, near the poles, the sunlight meets the Earth only glancingly, so that such heat as it contains is spread over a much greater area. There, even in summer, metal can get s cold that your flesh sticks to it, and if you expose your fing and toes for any length of time, the blood in them may free and you will lose them.

From rain forest to desert

The great heat around the equator has further consequence Hot air rises. It is also able to carry more water vapor than air. So above the equatorial lands there are great up-current moisture-laden air. But as they reach higher altitudes in the they cool and shed their moisture as torrential rain. As a res after millions of years of evolution, the equatorial lands of t Earth are blanketed with rain forest. Higher still in the sky, air currents, freed from their moisture, are deflected by the of the Earth beneath them and flow away on either side of t equator, north and south. In the slightly cooler latitudes, th sinks and at a lower level streams back toward the equator. Having lost its water, it now sucks up any moisture that rer on the land beneath and turns it into a desert. This global pattern of rain forest around the equator, with bands of par desert land to the north and south of it, has long been established, and today whole communities of plants and ani exploit these circumstances.

We are able to move from one environment to another

without major discomfort by simply putting on heavier or lighter clothing, or even taking it off altogether. But before we became globetrotters to the extent that we are today, we also became physically adapted to the particular climate in which we spent our lives. Those people living in the tropics acquired dark skins which protected them from the damaging ultraviolet elements in the strong sunshine. Others, in the Arctic, deprived of sun but afflicted with cold, developed permanent layers of fat just beneath their skin which kept them warm.

The spread of human beings from Africa, where they evolved, to all corners of the globe was achieved, as far as we can judge from the evidence currently available, during the past 300,000 years. That, in evolutionary terms, is relatively recent. Most of the species of animals and plants with which we share the world have been resident in their own particular habitats for much longer than that. In consequence, most have become much more closely matched to their particular environments than we have.

Bears, for example, are found throughout Eurasia and the Americas. Those in the north have become very big. Large size is an advantage in the cold, since a big body retains heat very much better than a small one. They have also developed thick shaggy fur to protect themselves from the cold. The species that lives farthest north, the polar bear, has in addition become white, so matching its surroundings of snow and ice, developed partially webbed toes to enable it to swim more efficiently between the ice floes, and pads of long coarse hair on its feet to insulate it from the cold. In contrast, the species of bear that lives in the warm humid forests of Malaysia is about half the length and a fifth the weight of a polar bear and has a very thin black coat.

Attuning to surroundings

But such specializations have one huge disadvantage. The more closely attuned a body becomes to its surroundings, the less able it is to survive elsewhere. We are the only species that can put on or shed a coat in a moment to compensate for a swift environmental change. A Malayan bear would die within hours in the Arctic; and a polar bear would collapse with heat-stroke in a tropical jungle. The same principle applies to plants, which are as little able to adapt to unfamiliar conditions as animals. All gardeners know that if they want to grow a plant that originated in another part of the world, they have to pay close attention to its particular requirements. The temperature, the humidity, the balance of light and shade—all have to be just right if a plant is to flourish.

Some species have to be protected from sunshine; others will only grow if they can bask in it. Some need frequent drenching with water, others, if treated that way, will die. All such requirements are simply a measure of how closely adapted a plant has become to the conditions of its original home.

Environmental changes

The climatic conditions in any one part of the Earth are not, however, eternal. Over geological time they have varied considerably. As we have seen, the continents themselves have shifted their position on the globe. Three hundred million years ago, the continent that was to become North America lay on the equator while southern Africa was much nearer to the South Pole.

There has been another great variable, too. For reasons discussed on p.30, the whole globe has warmed and cooled. The latest manifestation of such a change took place around 70,000 years ago. The world became colder. Ice fields and glaciers spread down from the poles and created a glacial period. After a series of oscillations in global temperature, that cold period finally came to an end around 10,000 years ago.

Although these environmental changes may seem extraordinarily swift when considered in terms of geological history, in the context of the lives of animals, they have for the most part been very gradual, being spread over many hundreds, if not thousands, of generations. Some species were able to move their territories, keeping pace with the shifting pattern of the climate. Others, through the processes of natural selection, changed physically and acquired new adaptations suited to the new conditions, which enabled them to survive. So ecological catastrophe has been avoided and the biological richness of the Earth has been maintained.

Patterns of
plant life

A habitat provides both a physical setting in which a plant or animal is able to survive and the requirements necessary for its growth. For an animal, these are its food in the form of plant material or prey animals. For plants, the requirements are sufficient light for photosynthesis and a supply of water and nutrient elements from the soil.

Since plants are static and often relatively large, they are an important component of any habitat. Their type and structure are determined by local climate conditions.

Raunkiaer's classification
The vegetation of the world may be broadly described in terms based on its general form and architecture—forest, grassland, scrub, for example. These terms are familiar, but they are difficult to define scientifically. In 1934 a Danish botanist, Christen Raunkiaer, developed a simple yet effective system for defining vegetation. He suggested that the response of vegetation to climate could best be judged by the height at which it held its perennating organs—that is, the buds, rhizomes, or bulbs by which the plant could survive through unfavorable weather conditions. Thus, in a favorable climate where buds are not in danger of drying or freezing, they can be held well above the ground; a complex, layered canopy of leaves then develops. Plants of this type, consisting of trees and tall shrubs, he termed phanerophytes.

Generally, the more difficult the climate, with periods of drought or cold, the lower the potential height of the canopy, for buds held at great height would be most exposed to climatic stress. In very cold conditions exposure to wind and ice eliminates all species with buds protruding above a cushionlike cover, usually less than 10 in in height. The dwarf shrubs and perennial herbs making up this cover are called chamaephytes.

Herbaceous plants have developed other ways of avoiding unfavorable conditions. Many plants in the temperate zone die back in winter and grow up again in summer. They bear their perennating buds at the soil surface, sometimes surrounded by a rosette of overwintering leaves (as in plantains), or with no leaves showing at all (as in stinging nettles). These Raunkiaer described as hemicryptophytes, meaning half-hidden.

Those plants that survive an unfavorable period by means of underground organs, such as bulbs or tubers, are termed geophytes. Another type of plant—the therophyte—survives periods of stress in the form of a seed. These annuals and ephemerals are an important part of the flora of deserts and semi-arid areas, and in other climates thrive as opportunists or weeds.

Evergreen plants
There are many other ways in which plants vary in response to climate. One is by having evergreen leaves. Evergreens occur in varied situations—the northern coniferous forests have many and yet so do the tropical rain forests. The evergreen leaf enables the plant to photosynthesize all year round in the tropics—or to take maximum advantage of a short growing season in colder climates. It is also economic in terms of the investment of energy and nutrients required for leaf construction. Making new leaves every year is costly to the plant, and nutrients are in short supply in both the rain forest and tundra soils.

The greatest disadvantage of being evergreen is that water is lost through the leaves at all times. This means that maintaining a leaf canopy puts plants under stress if water is scarce. The deciduous plant overcomes this by shedding its leaves in times of drought or when cold conditions make water scarce. Most evergreen leaves have adaptations, such as waxy surfaces, hairs, and sunken pores, that help reduce water loss.

Low temperatures and low water availability are the major climatic stresses that vegetation has to face. The structure of the world's vegetation is largely determined by the degree to which it is subjected to one or other of these stresses.

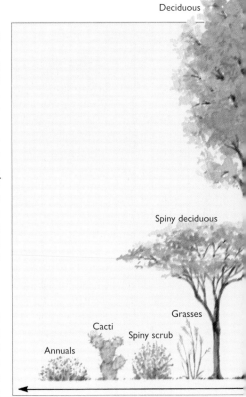

Deciduous

Spiny deciduous

Grasses

Cacti

Spiny scrub

Annuals

The benefits of being evergreen

Pines, such as the eastern white pine (*Pinus strobus*), succeed in temperate and boreal forests where the summer growing season is short. Because pines are evergreen they can start to photosynthesize and grow as soon as spring arrives.

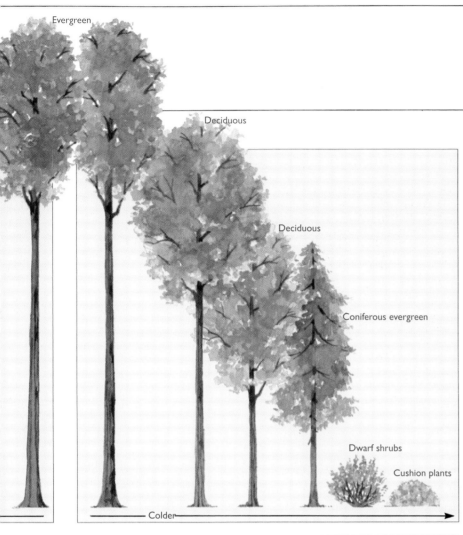

Evergreen

Deciduous

Deciduous

Coniferous evergreen

Dwarf shrubs

Cushion plants

Colder

The structure of a plant fits it for growth under particular climatic conditions. Tall, broad-leaved evergreens, for example, thrive only in warm moist conditions. Adaptations in structure and leaf form equip other types of plants for the rigors of colder or drier climates.

Each biome has its own characteristic spectrum of plant life forms. In tropical rain forest, for example, more than 90 percent of the flora are phanerophytes—tall trees and shrubs. In the Arctic tundra, some 60 percent of species are chamaephytes—dwarf shrubs and low-growing perennial herbs.

Certain plant forms are associated with particular climatic conditions. The cacti of deserts and semi-arid lands are the most familiar succulent plants, but succulence is also a characteristic of plants in other locally dry habitats. The stonecrops of alpine rock faces and the tree-dwelling cacti of the tropical rain forest are just two examples.

Many broad-leaved evergreen trees, such as *Symphonia*, grow in tropical rain forest. Since there is no dry period, plants do not need to shed their leaves to conserve water, and an evergreen leaf can survive several years.

In mediterranean climates, spring and fall, when conditions are warm and moist, are the best times for plant growth. The leaves of evergreens, such as the strawberry tree (*Arbutus unedo*), are ready to photosynthesize when these brief but favorable periods arrive.

53

Climate and plant life 1

The patterns of plant distribution on Earth are closely related to today's climate. Plants' tolerance varies, and because climate also varies in a complex way, plant distribution is complex.

Climate is governed to a great degree by the amount of energy received from the sun. Most solar energy reaches the surface of our planet at the equator, because here light energy has had to travel a shorter distance through the atmosphere than it has at the poles. Seasonal variations depend on the angle of the sun to the Earth, but the angle is always more oblique at the poles, so that a greater thickness of atmosphere must be crossed by the sun's radiant energy. At high latitudes, the low angle of the sun also means that the available energy is spread over a larger surface.

Atmospheric patterns

The uneven distribution of energy over the Earth causes instability in the atmosphere. In the cold polar regions the air is chilled and contracts. As it does so, it becomes more dense, descends toward the Earth's surface, and creates high pressures. Meanwhile, at the equator the air is heated from below, becomes less dense, and is pushed upward, which creates low atmospheric pressure systems.

However, the air that rises over the equator does not get as far as the poles before it falls again. It descends in a region around the Tropic of Cancer in the north and the Tropic of Capricorn in the south, creating two high-pressure belts. Some of this descending air moves poleward and collides with polar air heading toward the equator. This produces the unstable conditions typical of temperate regions.

From the global pattern of atmospheric circulation, it is possible to discern the direction of prevailing winds. However, the spin of the Earth—deflecting winds to the right in the northern and to the left in the southern hemisphere—also has an effect on direction. So, too, do mountain chains and local convection currents produced over large landmasses outside the equatorial regions (as they heat up in

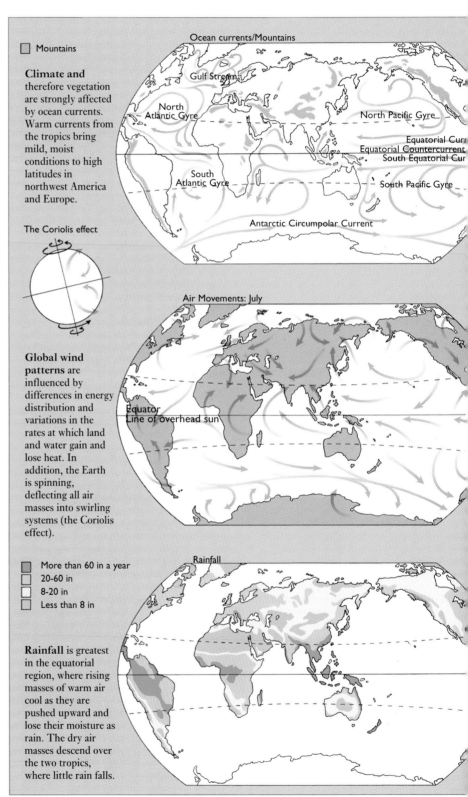

Climate and therefore vegetation are strongly affected by ocean currents. Warm currents from the tropics bring mild, moist conditions to high latitudes in northwest America and Europe.

The Coriolis effect

Global wind patterns are influenced by differences in energy distribution and variations in the rates at which land and water gain and lose heat. In addition, the Earth is spinning, deflecting all air masses into swirling systems (the Coriolis effect).

- More than 60 in a year
- 20-60 in
- 8-20 in
- Less than 8 in

Rainfall is greatest in the equatorial region, where rising masses of warm air cool as they are pushed upward and lose their moisture as rain. The dry air masses descend over the two tropics, where little rain falls.

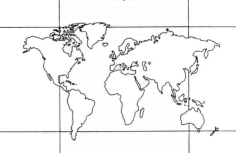

Equatorial regions receive the highest levels of energy input because the sun is directly overhead. This varies from season to season, however, with the tilt of the Earth on its axis (see diagram *below*).

In the northern hemisphere winter, the equatorial zone of wind convergence moves from north of the equator (see Air movements: July) farther south. This movement is caused by the Earth's tilt and shifting energy balance.

Plant productivity is dependent on warmth and moisture, so the highest levels occur in equatorial regions. Drought limits productivity in zones immediately to the north and south of these areas, and winter cold restricts the length of growing periods in higher latitudes.

Earth's energy balance

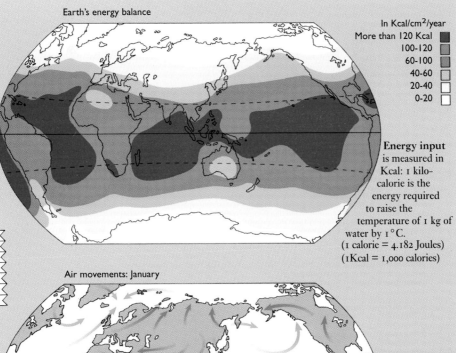

In Kcal/cm²/year
More than 120 Kcal
100-120
60-100
40-60
20-40
0-20

Seasonal energy change
July   January

Energy input is measured in Kcal: 1 kilo-calorie is the energy required to raise the temperature of 1 kg of water by 1°C. (1 calorie = 4.182 Joules) (1Kcal = 1,000 calories)

Air movements: January

Land productivity

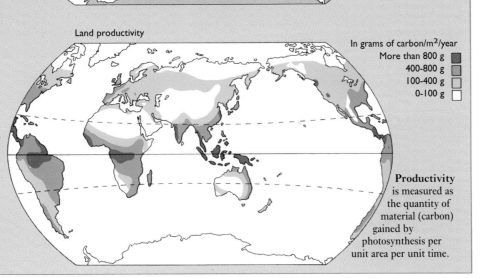

In grams of carbon/m²/year
More than 800 g
400-800 g
100-400 g
0-100 g

Productivity is measured as the quantity of material (carbon) gained by photosynthesis per unit area per unit time.

The greatest ever recorded fluctuation in temperature over 24 hours occurred in Browning, Montana in 1916. The temperature fell 100°F from 44°F to −56°F.

Rain falls on 335 to 350 days a year at Mount Waialeale on Kauai, one of the Hawaiian Islands, making it the most consistently wet place on Earth. Mean annual rainfall is 451 in.

Climate and plant life 2

summer and cool down in winter). The precise influence of prevailing winds also depends on the type of terrain they have crossed. Air masses that have traveled over the oceans, for example, are usually heavy with moisture.

Ocean influences

Because seawater warms up more slowly in summer and cools down more slowly in winter than the land, coastal regions have less extreme or "oceanic" climates, while interiors of landmasses have more extreme "continental" climates.

The circulation pattern of the oceans also influences climate. The warm Gulf Stream Drift of the North Atlantic, for example, insures mild, ice-free conditions in the summer months in the north of Scandinavia. But summer conditions are far more severe in the north of Alaska, at similar latitudes, where there is no such warm current.

Variable factors of atmosphere, wind, and ocean interact to determine the exact climate of any particular area. The equatorial zone has high levels of precipitation because water vapor in the rising air cools and condenses as rain. High-pressure areas, with descending air, have little rainfall. These are the desert zones—hot in the tropics and cold at the poles.

In the middle latitudes the collision of warm and cold air leads to atmospheric instability. This results in cyclones or depressions that form over the oceans and drift eastward, bringing periods of high precipitation to the western edges of the continents.

As the seasons change, the positions of the Earth's climate belts are altered. In the northern summer, for example, the tropical rain belt shifts north, pushing the other belts in front of it. Changing wind patterns take monsoon rains into southern Asia, while the Mediterranean, continental North America, Europe, and Asia experience dry seasons.

Plants and productivity

All the animals on Earth depend ultimately on plants for their energy. Because of

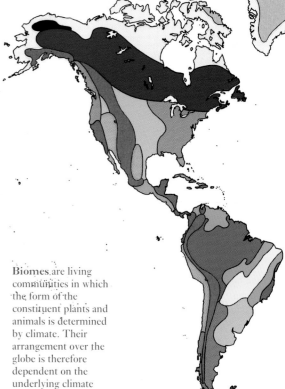

Biomes are living communities in which the form of the constituent plants and animals is determined by climate. Their arrangement over the globe is therefore dependent on the underlying climate pattern.

- Tropical rain forest
- Tropical seasonal forest
- Savanna
- Semi-arid scrub
- Desert
- Mediterranean
- Temperate grassland
- Temperate forest
- Boreal forest
- Tundra

this plants are described as primary producers. Even carnivores would be unable to make a living if there were no herbivores in the food chain. Plants depend for survival and growth on the energy of sunlight. This they capture, using the green pigment chlorophyll, and store in the form of energy-rich substances such as starch or oils.

The amount of new energy that is acquired by plant life in a given unit area of land in a given time is described as the

primary productivity of that unit. This reveals how much material (leaves, flowers, branches, trunk, and so on) is gained each year, and it can be measured in units of weight (such as pounds per square foot per year) or in units of energy (Joules per square foot per year).

Primary productivity influences both the number and the kind of animals that can live in a particular area. The quantity and variety are greatest where conditions are best for plant growth, but decline

The coldest place in the world is Vostock, Antarctica. It has a mean annual temperature of −72°F, and a record low of −128.6°F, registered on July 21, 1983.

The highest shade temperatures ever recorded are 136°F in Libya in 1922 and 134°F in Death Valley, California, in 1913.

The Atacama Desert in Chile is the world's driest place, with an annual mean rainfall of nil. In some areas there has been no rain in living memory.

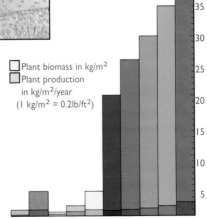

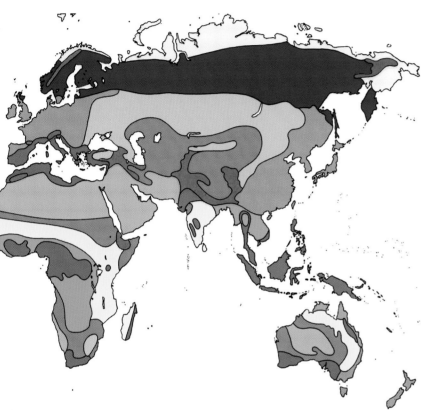

Ice
Mountains

under harsh conditions such as cold or drought.

Another important measure of habitat is the amount of living material actually present at a given time—the biomass. A large biomass usually means that both vegetation and animal life are diverse and complex. Plants often need to grow large in order to capture maximum amounts of sunlight. However, many leaves means great water loss, so drought can limit biomass development.

The amount of living material in a community is its biomass, measured in dry weight per unit area. Plant productivity is the rate at which new material is added to this by photosynthesis. This, too, is measured as dry weight, added within a specific area in a given time.

☐ Plant biomass in kg/m²
☐ Plant production in kg/m²/year
(1 kg/m² = 0.2lb/ft²)

45
40
35
30
25
20
15
10
5

Tropical rain forest

Life is found at its most luxuriant in the tropical rain forests. Plants grow taller and in greater profusion than in any other biome, and rain forests contain as many as 50 percent of the world's plant and animal species.

More of the sun's energy is fixed here by photosynthesizing plants than in any other biome, and so much carbon is contained within the timber that if it were all converted to carbon dioxide by forest destruction and burning, it would alter the climate of the whole Earth. Yet this apparently robust ecosystem is fragile, and disturbance soon upsets its complex but delicate balance.

Climate: rain forest is uniformly warm, usually between 68°F and 82°F year round. High rainfall is evenly spread throughout the year in this the wettest of all biomes; it always exceeds 60 in annually and may reach 400 in.

Structure: there are between three and five distinct layers of foliage in the forest, from the shrubby understorey to the giant trees that tower above the main canopy (see diagram, *top right*). This stratification gives strength to a forest structure in which the tallest trees may be 150 ft high.

Composition: rain forest is the most diverse of all biomes; up to 80 tree species per acre have been recorded in some areas, compared with 4–8 per acre in temperate forests. The high energy input from the sun means abundant resources for plant growth, which encourages diversity. The trees support rich communities of epiphytes and climbing plants, and the resulting complex structure provides many niches for animal species. (Epiphytic plants use other plants as a means of support and have no contact with the ground.)

Productivity: this is both the bulkiest and the most productive biome. Between 0.2–0.7 lb per sq ft is produced annually by a biomass of approximately 9 lb per sq ft (see pp. 56–57).

Animal life: each forest layer has its own community of animals; a species may remain in a particular layer for its entire life. Canopy dwellers include sloths, monkeys, lemurs, parrots, and snakes. On the ground there are peccaries, anteaters, tapirs, and capybaras. Rain forests are particularly rich in insect life, and in addition to those known there are millions of species as yet unidentified.

Nutrient cycling: rain forest soil is nutrient poor and highly leached and most of the available nutrients are in the plant biomass. Since temperature and humidity are so high, decomposition is fast and nutrients are rapidly cycled through the plant litter.

Conservation: major threats are from logging and clearance, the latter mainly for short-term agricultural exploitation, particularly for beef cattle. The resulting extinction of many species of plant and animal means the loss of many potentially useful organisms for agriculture or as sources of drugs.

Rain forest climate:
San Carlos de Rio Negro, Brazil

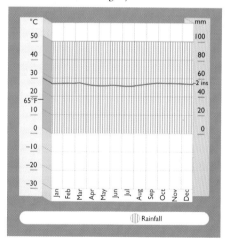

The richest environment on Earth is the tropical rain forest, where the lush plant life thrives on the constant high temperatures and abundant rainfall of the equatorial regions (*see above*). There is no drought, frost, or other seasonal change to disrupt growth.

Tropical
seasonal forest

The plants of equatorial rain forest have only to compete with one another for the resources around them. But a little farther from the equator a new environmental problem emerges—periodic drought. Even a short dry spell has a great impact on the life of the forest, from the trees down to the tiny invertebrates feeding among the leaf litter.

Climate: it is warm all year round, with plentiful rainfall that is limited to the wet season. The length of the dry season varies according to location. In the monsoon forests of northern India and Southeast Asia, it usually lasts from November until April. The drought interrupts plant growth and causes many species to lose their leaves.

Structure: the seasonal forest is less complex than the true rain forest and usually has only three vegetation layers. It merges with the rain forest on one side and the savanna woodlands on the other.

Composition: these forests are less diverse than the rain forest but still rich in species compared with temperate areas. Deciduous trees, such as teak and mountain ebony, are typical. There are also drought-tolerant evergreen species, such as the Indian banyan and the eucalyptus trees of northern Australia.

Since trees lose moisture through their leaves, deciduous species have an advantage in the tropical seasonal forest; they lose their leaves when water is scarce, in the dry season, and in this way cut down water loss.

Productivity: depending on total rainfall and the length of the wet season, productivity can be high. It is usually between 0.2–0.5 lb per sq ft a year with a biomass of about 7 lb per sq ft (see pp. 56–57).

Animal life: canopy dwellers include the langur monkeys of India, the koalas and cockatoos of Australia, and the widely distributed fruit bats and hornbills.

Ground dwellers, such as elephants and okapi, may browse upon the lower tree canopy, while various rodents, and ground-dwelling birds such as guineafowl and cassowary, feed on the forest floor.

Nutrient cycling: most of the nutrient reserve is in the plants themselves; little is in the soil. Although in the dry season a sudden supply of leaf litter is deposited on the soil surface, its decomposition is delayed until the arrival of the wet season. The rains can also cause extensive soil erosion.

Conservation: clearance for timber and overgrazing now pose serious threats to some seasonal forest areas such as those in northern India. Intensive management for wood production is destroying the ecological balance in other areas.

Where forest has been removed, the remaining vegetation is exposed to scorching drought during the dry season and most dies; rains erode the bare soils and flood the valleys, often with devastating effects on local human and animal populations.

Tropical seasonal forest climate: Belgaum, India

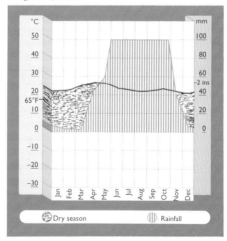

In the eucalyptus forests of Western Australia, a marked dry season (*see above*) disrupts plant growth. Some trees may lose their leaves, but others have tough, drought-resistant foliage. A period of high rainfall follows, but the temperature remains much the same year round.

Savanna

Vast open landscapes make these seasonally dry tropical grasslands among the most visually spectacular of all biomes. During the dry season the golden savanna appears parched, but, when the rains arrive, it quickly fills with green vegetation and flowers.

Savannas support an unusually high number of large animals, both grazers and predators. In the relative openness of their grassy habitat, these animals can easily be observed, and filmed, making this one of the most studied of all communities.

Climate: there is some variation in temperature, but the savanna is warm throughout the year. Dry and wet seasons alternate, often with quite severe drought during the dry season.

Structure: tall uniform grasses, often over 3 ft high, are the most conspicuous vegetation. Trees are scattered over the land with a density of sometimes 100 per 2.5 acres, although this is highly variable.

Composition: a wide range of grass species occurs in the savanna, but a given area is often dominated by just one or two species. These may have different growth periods and different degrees of tolerance to being grazed, so they can coexist.

Common trees in African savanna are the baobab, with its swollen water-storing trunk, and the acacias, with their characteristic flat tops.

Productivity: the production rate of between 0.04–0.4 lb per sq ft a year is achieved by a biomass of 0.04–1 lb per sq ft (see pp. 56–57), depending on tree density. Although savanna productivity is half that of the tropical rain forests, it achieves this with less than a tenth of the biomass; so it is, in effect, more efficient.

Animal life: herbivores, including antelope, zebra, and buffalo, feed on the savanna grasses. Trees are browsed by elephant and giraffe. The herds of large herbivores are preyed on by such carnivores as hunting dogs, hyenas, and lions.

Nutrient cycling: reserves of nutrients may be built up by the trees but, since both periodic fires and grazing keep tree density down, there is no great nutrient build-up in the savanna.

Some grasses secrete chemicals that inhibit microbe activity in the soil and reduce the rate at which nutrients are cycled and made available to other plants. This keeps the soil poor in nutrients and inhibits the growth of competitors.

Conservation: fire, which occurs naturally in the savanna, can also be artificially induced as an important tool in its maintenance; it can be used to prevent the development of forest since grasses regenerate after fire much more successfully than trees. Too frequent fires, however, can make the soil unstable by destroying all the surface vegetation and roots.

Grazing by herbivores actually enhances productivity by stimulating growth, but the density of animals has to be controlled. Overintensive grazing by cattle, however, can cause the encroachment of scrub vegetation onto savanna.

Savanna climate: Harare, Zimbabwe

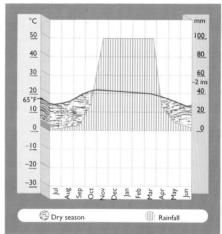

Dry season   Rainfall

Scattered trees among a sea of grass characterize the East African savanna. Herds of large herbivorous animals thrive on the lush grazing, but in the dry season the grass dries up and water becomes scarce. Rainfall is restricted to the summer months (*see above*).

Semi-arid scrub

Where rain is infrequent, or limited to certain short periods of the year, where summers are so hot that every last drop of moisture evaporates, coarse thorny scrub develops. This vegetation provides meager fare for the few animals able to feed upon it. Even humans can only scrape a living here by using hardy breeds of sheep and goat as pasture animals.

Climate: semi-arid land has an annual rainfall of 10–20 in. Summer temperatures are high, and any available moisture evaporates quickly. Winters may be cold; in the scrublands of the Middle East, rain and snow fall in winter.

Structure: woody shrubs up to 6.5 ft high are the main vegetation. If the land is not too heavily grazed, they are sometimes interspersed with dwarf shrubs and grasses. The larger bushes are usually relatively widely and evenly spaced. This even spacing results from intense competition for the limited resources, particularly water—the shrubs extend their root systems over a wide zone to obtain sufficient water and young shrubs cannot establish themselves near them.

Composition: the shrubs in this biome are often sclerophylls—plants with small leathery leaves, well adapted to drought. The deciduous creosote bush family is abundant, as are the tamarisks, especially in dried-up water courses. Bulb plants, such as desert lilies and desert hyacinths, are widespread, as are annuals that survive drought periods as seeds.

Productivity: annual productivity in desert scrub is about 0.16–0.2 lb per sq ft (see pp. 56–57). If the land is heavily grazed, however, it may become virtual desert, and biomass and productivity fall.

Animal life: small herbivores, such as gerbils and hares, manage to survive in desert scrub, but large grazers, such as wild ass and sheep, are rare. Gazelles are perhaps the most common large animals. There are some carnivores, including pumas, leopards, and cheetahs, that prey on the grazers, and lizards and snakes abound.

Nutrient cycling: since these soils have a low organic content, they do not retain nutrients. Leaf litter is important in recycling the nutrients that do exist but it is easily blown away by the wind. The growth of plants is slow, however, so demand for soil nutrients is not high.

Conservation: while arable farming is difficult in semi-arid scrublands, pastoral farming is possible, and these areas are often overgrazed by domesticated sheep and goats.

The first result of overgrazing is that the most palatable and digestible plants, grasses, for example, are destroyed, followed by the more acceptable shrubs. Goats, which have a wider tolerance than sheep, then consume all but the more toxic plants, such as the joint pines (*Ephedra*). Eventually the land may become true desert. The state of the vegetation is an indication of how far this degeneration has progressed.

Semi-arid scrub climate: Astrakhan, Russia

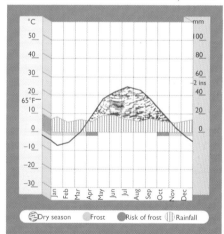

Dry season   Frost   Risk of frost   Rainfall

The Argentinian semi-arid scrub is typically sparse and low-growing, and contains only a few plant species. The shrubs provide some shelter for small animals such as rodents and snakes. There is a long, hot, dry season but winters can be cold and frosty.

Desert

Life itself is at the limits of endurance in the desolate, barren desert lands, where rainfall is so low that little vegetation can survive. Those living things that do exist in this biome must adapt to cope with the formidable conditions imposed by almost constant drought. Even human beings find it hard to survive the rigors of desert life.

Climate: deserts have less than 4 in of rain annually and may go for years with no rain at all. Some are hot all year round; others, such as the Asian Gobi Desert, are cold in winter. The hot deserts lie in the subtropical high pressure belt, while those with cold winters are at high altitudes and in the shadow of mountain ranges that prevent rain-bearing clouds from reaching them.

Structure: there is little vegetation and much open ground. What plant life there is may be restricted to wadis (old river beds) and is of vital importance in creating shaded microclimates for animals.

Composition: plants must be drought-resistant. Some, such as the Old World euphorbias, have small, thick leaves that are short-lived and can be shed in the driest periods. The cacti of the American deserts store water in their bulky stems. Many desert plants open their pores at night instead of in daylight and fix carbon dioxide while temperatures and evaporation rates are not so high. A plant of the Namib Desert, *Welwitschia*, actually absorbs nighttime mist through its leaf pores.

Some plants, such as members of the gourd family, have deep roots to seek out water far below the ground, but cacti and others root in the surface layers of the soil and take up the dew. Annuals survive long droughts as seed and go through a rapid life cycle during the rare rains.

Productivity: desert productivity is usually under 0.06 lb per sq ft a year (see pp. 56–57), a similar rate to that of the tundra and open oceans.

Animal life: reptiles are common, and desert species adapt to the conditions by excreting almost dry, crystalline urine to conserve water. Small mammals, such as gerbils and ground squirrels, survive by burrowing to avoid the heat, but there are also larger hardy grazing animals such as gazelle and oryx. Successful predators include foxes, hyenas, and some cats.

Nutrient cycling: the nutrient cycle is almost nonexistent. Because of the lack of organic matter in the soil, there is little bacterial activity. Decomposition is slow and the few nutrients take a long time to recycle through the ecosystem.

Conservation: deserts are currently spreading for two reasons: the subtropical drought resulting from climatic change (see pp. 156–57), and human mismanagement. For example, the remaining vegetation of the deserts is at risk because of overgrazing by domestic stock, particularly camels and goats. Many wild desert animals and birds, such as the Arabian oryx and the houbara bustard, are seriously threatened by hunting.

Desert climate: Aswan, Egypt

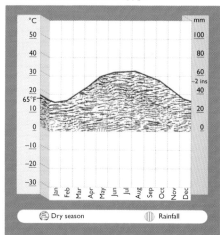

Dry season   Rainfall

Intense heat, prolonged drought, and mobile sandy soils make desert areas such as these in Saudi Arabia almost uninhabitable. Rainfall may be irregular, even virtually nonexistent, but dew can supply the needs of some hardy plants and animals.

Chaparral lands

Areas with a mediterranean climate—mild winters and hot summers—have always been attractive to humankind. This biome not only fringes the Mediterranean Sea, it also occurs in the southern tip of South Africa, southern Australia, and California, where it is known as chaparral.

Climate: typically, winters are cool and moist, with the possibility of frost, while summers are hot and dry. The summer drought is the most unfavorable period for plant and animal life.

Structure: many of these areas, including the Mediterranean region itself, were originally covered by woodland. Fragments of woodland remain, but much of this biome, particularly in the Mediterranean area, has been degraded to scrub (maquis) or even to low heath or open grassland (garrigue) as a result of human activity and the grazing of domestic stock.

Composition: trees include both evergreen species, such as live oaks, the strawberry tree (*Arbutus unedo*), and eucalyptus, and deciduous trees, such as the lombardy poplar. Conifers, such as junipers and pines, are also characteristic.

The dwarf shrubs of mediterranean heathland include many aromatic species, lavenders, sage, California lilacs (Ceanothus), and plants with attractive flowers such as rockroses (Cistaceae) and mesembryanthemum.

Among the grass species of the Mediterranean are the wild wheats and barleys that were among the first plants to be domesticated 10,000 years ago.

Productivity: in wooded regions productivity may reach 0.3 lb per sq ft annually. In poor garrigue, however, it may be as low as 0.04 lb per sq ft (see pp. 56–57).

Animal life: mediterranean vegetation is grazed by animals such as the red deer and wild boar in Europe, gray kangaroo in Australia, duiker and hyrax in South Africa, guanaco in Chile, and mule deer in California. Predators include wolves and coyotes.

Nutrient cycling: in woodland areas much of the nutrient reservoir is held in the plants, and the alluvial soils are rich. But in heath and scrub the soil retains little nutrient content and may easily become further degraded.

Conservation: for centuries, the mediterranean lands have been intensively exploited by human populations and used for grazing sheep and goats and other animals. This has destroyed much of the original forest, reducing plant biomass and productivity and resulting in extensive soil erosion.

In all mediterranean climate areas, fire is an important ecological factor. The hot, dry summers and the volatile oils produced by many plants render the vegetation highly flammable. Most plants can survive fire and some are even stimulated into growth by it. Human management of this biome needs, therefore, to permit occasional fires.

**Mediterranean climate:
Pasadena, California**

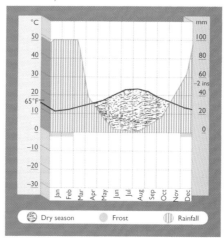

Dry season   Frost   Rainfall

Aromatic, spring-flowering shrubs and brightly colored flowering bulbs bloom in the scrubby heathland of Crete. The evergreen woodland that once grew here has been degraded by centuries of grazing. Summers are hot and dry, winters cool and wet.

Temperate grassland

The North American prairie, South American pampas, Eurasian steppe, and South African veld are all temperate grasslands. Landscapes may appear monotonous and lacking in relief—hot summers leave the grasses dry and brown and the trees are confined to the riversides—but the grasslands have proved valuable to humankind. The many grazing animals have long been hunted and the rich soils have provided excellent agricultural land. The opportunities have not always been well used, however, and some grassland has been destroyed.

Climate: grasslands occur in the interiors of landmasses where continental climatic conditions bring cold winters with hard frosts, and hot, dry summers. Summer drought, occasional fires, and intense grazing are all factors that prevent the growth of trees.

Structure: tall-grass prairies may be more than 6.5 ft high while short-grass areas are less than 2 ft. The grass shoots grow densely and trap much of the available light, leaving little for other plants. Trees are rare and confined mainly to damper areas in the grassland where they can obtain some water in summer.

Composition: grass species, such as fescues, bromes, and bluegrass, are typical, but many herbs also grow among them. The plants have characteristic times of growth: those that need more water grow earlier in the season, while the drought-tolerant species grow later.

Productivity: like savannas, temperate grasslands are surprisingly productive for their low level of biomass. An annual productivity of 0.1–0.3 lb per sq ft is typical, depending on moisture levels (see pp. 56–57).

Animal life: abundant low-growing vegetation means plenty of food for grazing animals. Large grazers include deer, saiga antelope, pronghorn, gophers, prairie dogs, and bison. Invertebrate grazers, such as grasshoppers and caterpillars, are also important members of the community and provide food for the many insectivores such as sharp-tailed grouse and prairie chickens.

Nutrient cycling: this occurs relatively rapidly through the decomposition of thick black humus in the surface layer of soil. Despite the many grazing animals, about 90 percent of the plant productivity is left to die and decompose, so litter is plentiful and nutrients are quickly released.

Conservation: the numbers of natural wild grazers are well balanced with grassland productivity, but when domestic stock is introduced it can create higher grazing pressures than the land can sustain. The vegetation cover is then reduced, and only plants that are unpalatable to the grazers survive.

Arable farming has proved successful on the fertile soils, but overuse can lead to soil instability and wind erosion in the dry summer.

**Temperate grassland climate:
Chkalov, Russia**

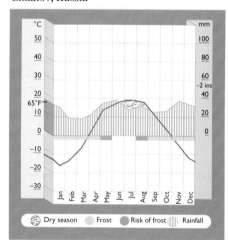

Dry season — Frost — Risk of frost — Rainfall

Herds of bison graze in the prairie grasslands of Yellowstone National Park. Grasses are the dominant vegetation and there are few trees except along riversides. Rainfall is adequate year round, but there is a short dry season in summer when temperatures are high.

70

Temperate forest

Much more of the temperate world would still be clothed in forest if it were not for a long history of clearance for agriculture and settlement. This biome now includes some of the most densely populated areas of the world.

Climate: rainfall in this biome is always adequate for tree growth and is usually fairly evenly spread throughout the year. In some localities, such as eastern Asia, southeast Australia, and the Pacific northwest, rainfall is so high that the luxuriant forests are called "temperate rain forests."

Summers are warm, but winters may be cold, often falling below freezing.

Structure: temperate forests are simpler in structure than the tropical jungles. There are usually only two canopy layers—the trees and an understory of shrubs. Sufficient light penetrates through the canopy to the forest floor to allow the development of an herbaceous and a moss ground layer.

Composition: there are several different types of temperate forest. Coniferous forest (Pacific coast of North America) contains redwood, hemlock, and western red cedar. Mixed conifer and deciduous (Great Lakes region) has oak, birch, hemlock, pine, and maple.

Broad-leaved deciduous forest (eastern North America, western Europe) is also rich in oak species, together with beech, ash, and chestnut. The broad-leaved evergreen forests of Japan and Tasmania have such trees as chinquapin and southern beech.

Compared with tropical forests there are relatively few tree species in temperate forests, but there is a rich ground flora. These are mainly hemi-cryptophytes—perennial plants that survive the winter by dying back to ground level buds and then surge back into growth in spring (see pp. 52–53).

Productivity: an average biomass of 6 lb per sq ft produces some 0.12–0.5 lb per sq ft annually (see pp. 56–57).

Animal life: ground-dwelling creatures include badgers, bears, elk, wild pigs, woodchucks, and many insectivores and rodents, as well as predators such as wolves, wild cats, and foxes. The forest canopy houses squirrels and many bird species, particularly titmice, chickadees, woodpeckers, warblers, and finches.

Nutrient cycling: the main nutrient reservoirs are in both the plants and the soil. The leaf litter carries nutrients from the leaf canopy to the soil, but it may take several years for the organic matter to decompose completely and for the nutrients to be released. Rain passing through the canopy leaches out nutrients and carries them to the ground, and almost as much is cycled in this way as by leaf fall.

Conservation: agricultural clearance has left little virgin temperate forest, and soil degradation following clearance can lead to the formation of heathland and moorland. The fragments of ancient forest that do survive are distinguished by their diverse flora and fauna and the presence of species that are sensitive to disturbance.

Temperate forest climate: Luxembourg

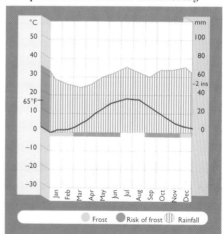

Glowing colors suffuse the temperate forest of Vermont in fall. Winter frosts mean water is in short supply and trees cannot afford the luxury of keeping leaves. As the green pigment, chlorophyll, breaks down, the leaves turn golden, red, or brown before they fall.

Boreal forest

The forests of the far north form an almost continuous belt around the globe, across northern North America and Eurasia. The growing season in these high latitudes is too short for most deciduous trees (apart from birches) to develop a full leaf canopy, but evergreens, able to start photosynthesis at the first sign of spring, can survive. Unlike the broad-leaved evergreens of farther south, however, these conifers have tiny, needle-shaped waxy leaves that can withstand the intense cold and drought of winter.

Climate: there is great seasonal variation, with hot summers and extremely cold winters. In the eastern Siberian larch forests, the variation in extreme annual temperatures can be as much as 180°F. Rainfall is not high, usually about 16–24 in annually; in winter it falls as snow. In the most northerly areas rainfall may be as low as 6 in, not much more than in desert regions, but since temperatures and evaporation rates are low, supplies are adequate.

Structure: the canopy is dense, allowing little light to reach the forest floor. Some dwarf evergreen shrubs may grow; beneath them lichens and mosses form ground cover. There are many bogs in low-lying, waterlogged areas.

Composition: needle-leaved trees predominate. Most, such as spruce, pine, and fir, are evergreen but some, larch and tamarak for example, are deciduous. A few broad-leaved deciduous trees, including birches, rowan (in Eurasia), alders, and willows, are hardy enough to survive.

The heather family, including blueberries and mountain heathers, is the most abundant in the understory. Marshland species such as labrador tea, bog rosemary, and leatherleaf grow on bog surfaces.

Productivity: the short growing season results in a lower productivity than that of the temperate forests. It generally ranges from 0.04–0.3 lb per sq ft a year (see pp. 56–57).

Animal life: the many insectivorous birds that live in the forests in summer migrate south in winter, but the crossbills remain year round, feeding on conifer seeds. Moose and reindeer are common, and there are many small mammals such as hares and voles. Predators include wolves, lynx, and owls.

Nutrient cycling: the boreal forest system is poor in nutrients. Therefore, when forests are being cropped for timber, it is important to replace nutrients by fertilizing. Lichens, which fix nitrogen from the atmosphere, play an important part in the replenishment of nutrients.

Conservation: many areas of boreal forest have been heavily exploited for forestry and timber production. In Finland, for example, little virgin spruce forest remains and most of the country is covered with even-aged plantations. Boreal forest areas also suffer particularly badly from the effects of air pollution and acid rain (see pp. 164–65). Their acid soil cannot neutralize the acidity in the rain and it passes on into lakes and rivers.

Boreal forest climate: Moscow, Russia

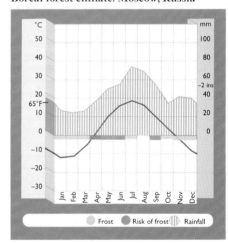

Evergreen coniferous trees predominate in the boreal forest of Banff National Park in Canada, but dwarf shrubs such as blueberry cover the ground. Summers here are warm, but winters are very cold and the growing season short. Frost is a risk for much of the year (*see above*).

Tundra

The tundra, a polar desert, lies beyond the limits of forest growth. In these high latitudes close to the Arctic, winter winds laden with crystals of ice blast all vegetation projecting above a cushioned carpet of ground-covering plants.

Despite months of total darkness in winter, these barren wastes manage to support some life. And in summer the reemerging dormant plants and insects are joined by migrants, the herds of animals and flocks of birds that have wintered farther south. This fruitful summer tundra provides rich pickings for residents and visitors alike.

Climate: winter temperatures are extremely low, often averaging –4°F to –22°F in the coldest months. High atmospheric pressure means that the chilled air carries little rain, but since evaporation rates are also low, supplies are adequate for most areas. The summer lasts three to four months and even then at the highest latitudes the average monthly temperature may not rise above freezing.

Structure: there are no true trees—they could not tolerate the winter wind and ice damage. Dwarf shrubs, sedges, grasses, and lichens, along with cushion-forming plants, form a low carpet of vegetation.

Composition: dwarf birch, several species of dwarf willow, and many other woody dwarf shrubs such as bilberry and crowberry form the basic vegetation. Tussocky grasses and cushion plants such as saxifrages are also an important feature.

Productivity: low temperatures combined with a short growing season result in an annual productivity of only 0.02–0.08 lb per sq ft, similar to that of desert.

Animal life: polar bears, musk oxen, and arctic foxes are among the resident animals that have to cope with food scarcity as well as extreme cold in the long winter. Lemmings, also residents, are able to continue grazing on vegetation beneath the snow cover. Tundra animals do not usually hibernate since the summer is too short for them to have built adequate fat reserves.

Mosquitoes, blackflies, and other insects overwinter as eggs, hatching in their millions as the temperature rises in summer. Birds migrate to the tundra from wintering grounds farther south to take advantage of this plentiful food source. Caribou herds also arrive to feed on the summer vegetation, bringing their predators, wolves, with them.

Nutrient cycling: the nutrient cycle is slow because there is little bacterial activity in the cold soil. Slow decomposition leads to the accumulation of peat and surface humus where nutrients may become locked in. But since growth is so limited by the cold, nutrients are not needed in large quantities.

Conservation: the exploitation of oil reserves in the tundra causes some serious environmental problems. Pressure of traffic damages soil, and animal migration routes are disrupted by aboveground pipelines.

Tundra climate: Fort Yukon, Alaska

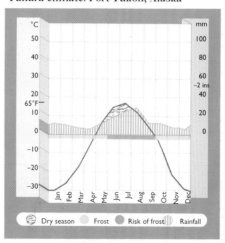

Lush vegetation flourishes in the long days of summer in the Yukon tundra, but the harsh dark winter, with its ice-laden winds, prevents the growth of any tall trees. Rainfall is low throughout the year and drought is possible on well-drained sites in summer (*see above*).

From valley
to mountain peak

Colorful plant life thrives below the towering bare summits of these Colorado mountains.

Going up a mountain is, in biological terms, rather like undertaking a much longer journey toward the poles: the changing vegetation types encountered with increasing altitude resemble the succession of biomes with increasing latitude. Mountains have a climate of their own and thus a distinctive flora and fauna, and add diversity to a region.

Mountain climates
Temperatures generally decline the higher above sea level one goes. The rate at which they fall is variable, but an increase in altitude of 300 ft results in a drop in temperature of approximately one degree Fahrenheit. The average annual temperature is always lower at higher altitudes, and so the growing season is shorter. Plants cannot begin their growth below certain limiting temperatures that vary from one species to another.

Temperature affects the degree of moisture retention of the atmosphere: cold air cannot hold as much moisture as warm. When air is pushed up a mountainside, it is forced to lose some water as precipitation, either rain or snow, depending on the altitude. High areas, then, are likely to receive increased precipitation, especially if they intercept warm, moist air that has been carried over the sea. On the highest mountains, most precipitation falls as snow; if summer temperatures are so low that it never melts, the snow may compact to ice, and a glacier is formed.

On the leeward side of a mountain ridge, however, where air is descending, rainfall is much scarcer, and such regions are said to be in a rain shadow. The Gobi Desert in Asia, for example, is in the rain shadow of the Himalaya Mountains.

Vegetation zones
Biomes around the world merge into one another, but the zones of vegetation on mountains are quite distinct. The plant species that occupy particular belts of vegetation vary from one mountain to another, but the structure of the belts is similar, even for widely separated mountain regions.

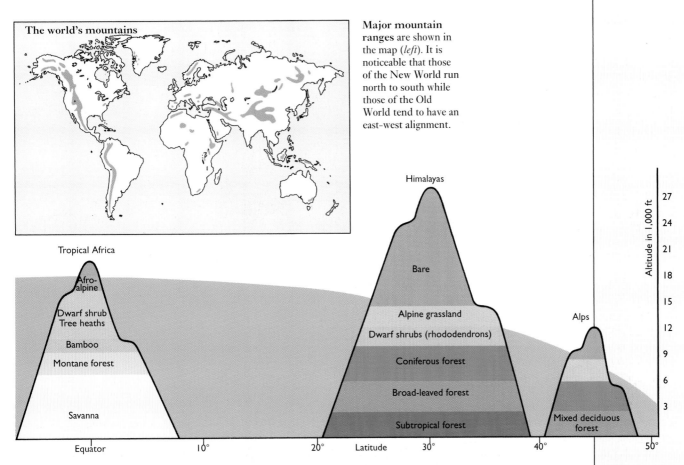

The world's mountains

Major mountain ranges are shown in the map (*left*). It is noticeable that those of the New World run north to south while those of the Old World tend to have an east-west alignment.

Tropical Africa

Afro-alpine

Dwarf shrub
Tree heaths

Bamboo

Montane forest

Savanna

Himalayas

Bare

Alpine grassland

Dwarf shrubs (rhododendrons)

Coniferous forest

Broad-leaved forest

Subtropical forest

Alps

Mixed deciduous forest

Altitude in 1,000 ft

27 24 21 18 15 12 9 6 3

Equator 10° 20° Latitude 30° 40° 50°

The Alps of Europe and the Himalayas of Asia, for example, both have a deciduous temperate forest zone followed by coniferous evergreen, dwarf shrub, then grassland zones before the bare scree slopes and ice take over. Both deciduous belts contain oaks, but the coniferous zone of the Himalayas is dominated by deodar cedar, whereas in the Alps spruce and larch are the most important trees. Rhododendrons and junipers grow in the dwarf shrub zone of both areas, which also share some alpine grassland plants.

The altitude at which the zone boundaries occur differs greatly. The coniferous forest zone, for example, stops at about 6,550 ft in the Alps but continues to about 11,150 ft in the Himalayas. This is partly due to the more southerly latitude of the Himalayas—they receive greater energy from the sun—and partly to warm air moving up from the Indian subcontinent.

The mountains of western North America have a similar pattern of zonation. In the Sierra Nevada, however, the temperate forest zone is not deciduous but consists of coniferous redwoods. Above this zone grow the boreal forests of red fir, lodgepole pine, and hemlock.

Some mountains, such as Mount Kenya, lie almost on the equator. Here the zonation pattern is somewhat different. Below the alpine grassland and scrub zones are bamboo and montane forest bands, replacing the boreal and temperate forest zones of the higher latitude mountains.

Just as there are fewer plant and animal species in high latitudes than in the tropics, so the diversity of species reduces with altitude where climatic stresses are greater. These result in less vegetation, lower productivity, and fewer opportunities for animals to make a living.

On a mountain, climate and plant life gradually change with altitude. The different vegetation zones that result correspond, on a smaller scale, to the biome changes that occur with the progression toward higher latitudes.

Hence the Himalaya Mountains have zones of deciduous forest and tundra similar to the biomes of regions much farther north. And mountains on the equator have zones of alpine vegetation near the summit.

Freshwater wetlands 1

The world's freshwater habitats provide a set of niche opportunities for aquatic animals and plants quite distinct from those of the oceans. Apart from being, typically, less salty, freshwaters differ from the sea in that they are divided into static ponds and lakes, and flowing streams and rivers.

Through a succession of infilling stages, highly productive static freshwaters often become converted into vegetation-covered land; and several of the stages in this conversion, such as swamp and marsh (see pp. 82–83), form characteristic types of freshwater wetland in their own right.

The conversion of a lake by way of marsh into woodland is just one example of the range of temporal changes that affect freshwater habitats. Streams and lakes silt up, rivers change their courses, and new lakes are formed by erosion or climatic change. The life of such a new lake will pass through several phases of colonization before a stable, complex community of plants and animals is formed.

The permanence of a freshwater habitat frequently depends on its size. A tiny ditch in a meadow or a wet-season pool in dry grassland is likely to be temporary and will probably not exist long enough to develop a complex living community. On the other hand, major deep lakes, such as Lake Baikal in Russia and Lake Victoria in Africa, are silting only slowly and have been in existence long enough to have developed their own unique faunas of species that occur only there.

Essentially, ponds and lakes differ only in size. Water enters them directly as rain or as streams and rivers flowing in from their catchment areas. The water carries sediment and dissolved minerals from the underlying and surrounding earth, and if the lake is shallow and productive, it will rapidly be converted into dry land.

The life of a lake is also influenced by the chemical content of its water. The main impact of this is in the effect that different levels of mineral salts have on the primary production of aquatic plants, including plankton. There is also a direct

The complex nutrient flow in a temperate lake is shown in this simplified food web diagram. Decaying detritus as well as algae and aquatic plants form the base of such a web. Herons and pike are examples of top predators.

effect on some invertebrate animals.

In terms of water chemistry, there are three types of lakes. Eutrophic lakes are usually shallow; water flows into them from nonacid soils and they have high levels of dissolved calcium, nitrates, and phosphates—features that make them very productive. The plankton and other water plants have high photosynthetic levels, while the presence of large amounts of calcium provides good growing conditions for mollusks and crustaceans. Since it contains high densities of plankton, the water is often greenish and opaque. Although oxygen levels may be reasonable in surface waters, near the organic sediment of the bottom bacterial decomposition leads to a shortage of oxygen.

Deep, clear, oligotrophic lakes are formed in landscapes that do not contain

elements such as calcium. The low levels of mineral salts in the water do not favor plant growth, and the lack of decaying organic material on the lake bottom, which consists of mineral sands or silt, means that oxygen is usually plentiful at all levels. Although the total amount of plant and animal life in an oligotrophic lake is often small, the number of different species may be considerable.

In dystrophic lakes, which only exist in habitats with extremely acid soil such as peatlands, productivity, total biomass, and species diversity are all low.

The broad common patterns of nutrient flow in eutrophic and oligotrophic lakes are not unlike those of the sea. Primary production rests with the plant plankton eaten by small animal plankton, which in turn form the food of other

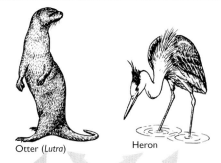

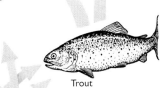

In flowing water aquatic plants are less important than in lakes. This food web shows some of the main types of animals in the river community. Nutrient flow in such a web moves from tiny invertebrates such as insect larvae to large predators such as otters and heron.

Otter (*Lutra*)

Heron

Eel

Trout

Dipper (*Cinclus*)

Stickleback

Shrimp

Freshwater snail

Insect larva

swimming animals such as fish.

Similar community patterns exist in the slower-moving, broad stretches of lowland rivers. Here, however, because of the one-way flow of the water, planktonic life is rare except in the slowest reaches. Much larger differences exist in the fast-moving upland sections of rivers, where the sheer physical force of the water means that there can be no stable sediment bottom to the habitat. In this swift-flowing world, most animals (apart from powerful swimmers such as fish) and plants must attach themselves to the river bottom or find sheltered spots in crevices under rocks or among attached vegetation. Primary production comes partly from these attached plants but also from vegetation that falls into the water from the banks.

Leaves falling from the bordering woodland provide a river with an important input of nutrients. Such plant litter provides energy for the detritus-consuming organisms at the base of the food chain.

81

Freshwater wetlands 2

The freshwater wetlands, unlike other major biomes, are not zoned over the Earth in a distinct pattern. Wetlands can occur wherever rainfall is abundant, or where temperatures are low, and evaporation slow, sometimes even on slopes and hilltops. Where rainfall is low and evaporation high, wetlands are usually restricted to basins that receive drainage water.

Plants of the wetlands

Wetlands are constantly changing, sometimes slowly, sometimes rapidly. An open body of water, with time, becomes silted up and is invaded by aquatic plants. These may be submerged types, such as the hornwort, or floating plants such as water lilies. Eventually, emergent species like reeds can colonize the accumulating mud.

The growth of the plants themselves actually encourages the silting up process, because it slows the flow of water and adds new organic matter to the accumulating muds. The result is a marsh—a sizable population of emergent water-dwelling herbaceous plants with a water table that is permanently high. Because the water table is above the sediment, even in summer, standing water is always present. Sometimes, however, the thick rhizomes of the plants, which are full of air channels to carry oxygen to the waterlogged roots, float in a carpet on the water surface.

In marshes of this sort, one plant species often dominates. In temperate areas this may be reed, sawgrass, or cattail (reedmace); in tropical swamps, papyrus. Trees, such as the swamp cypress of the Florida Everglades, may grow in the swamp with their trunk bases permanently under water, but often the surface water is populated by floating aquatics such as the water fern and duckweeds.

Swamps are among the most productive habitats on Earth, even exceeding the rain forests in their productivity, and they thus support a wide range of animal life. Aquatic invertebrates and fish abound in the standing water between the plants and feed on the detritus that falls from above.

Reedbed birds, such as the marsh wren, inhabit forests of vertical reed stems, feeding on insects and weaving their nests into the herbaceous scaffolding. Coots and bitterns also live in reedbeds; snakes and alligators are other vertebrate water-dwellers. Flying predators include the marsh harrier and Everglade kite.

As silts and peats accumulate in the swamp, the surface of the sediment rises in relation to the water table, so that in summer water may be invisible above ground. In such a habitat, known as a fen,

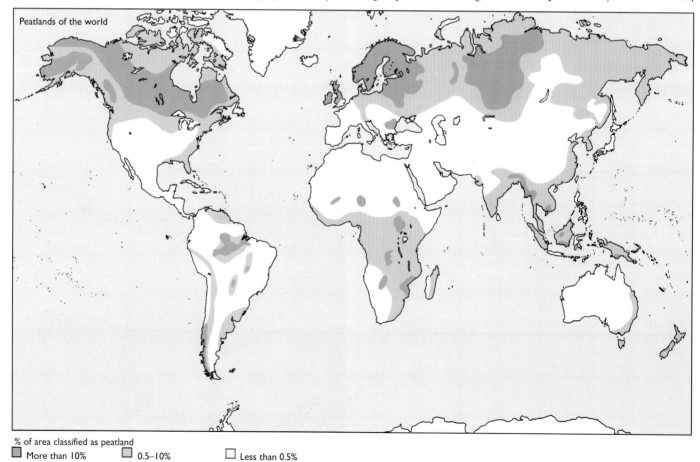

Peatlands of the world

% of area classified as peatland
■ More than 10%   □ 0.5–10%   □ Less than 0.5%

wet woodland trees may flourish.

How bogs are created

In high latitude regions of the world where precipitation is high but evaporation is low, this succession does not stop here. The bog moss, sphagnum, invades the habitat and alters the entire course of events. Sphagnum has some remarkable properties. It acts like a sponge and can hold 10 times its own weight in water. It attracts nutrients to its surface and binds them so that they are no longer available to other plants, thus leaving the wetland water acid and poor in nutrients.

As the sphagnum builds up its hummocks, a number of other species, such as the swamp trees, willow and alder, are killed off. Only acid-tolerant plants such as cranberries, sundews, and cotton sedges survive in this habitat, which is now known as a bog. Another way in which bog is created is if a floating carpet of sedges forms in open water and becomes the vehicle for sphagnum to invade.

Many wading birds breed in boreal bogs, including golden plovers, broad-billed sandpipers, greenshanks, curlews, and cranes, but because the productivity of these bogs is much lower than that of swamps, there is, by comparison, less diversity of species.

The peat of the wetlands may have taken more than 5,000 years to form. But the usefulness of peat as an energy source and soil conditioner threatens these ancient bogs with extinction—almost all the once-luxuriant bogs of central Ireland have been excavated and destroyed. Forestry is a further threat, since many conifers grow well on peatlands after the surface has been plowed and the water table lowered.

Peatlands, areas where peats have gradually accumulated on wetlands, are most frequent in cool temperate and boreal regions.

Wetland succession

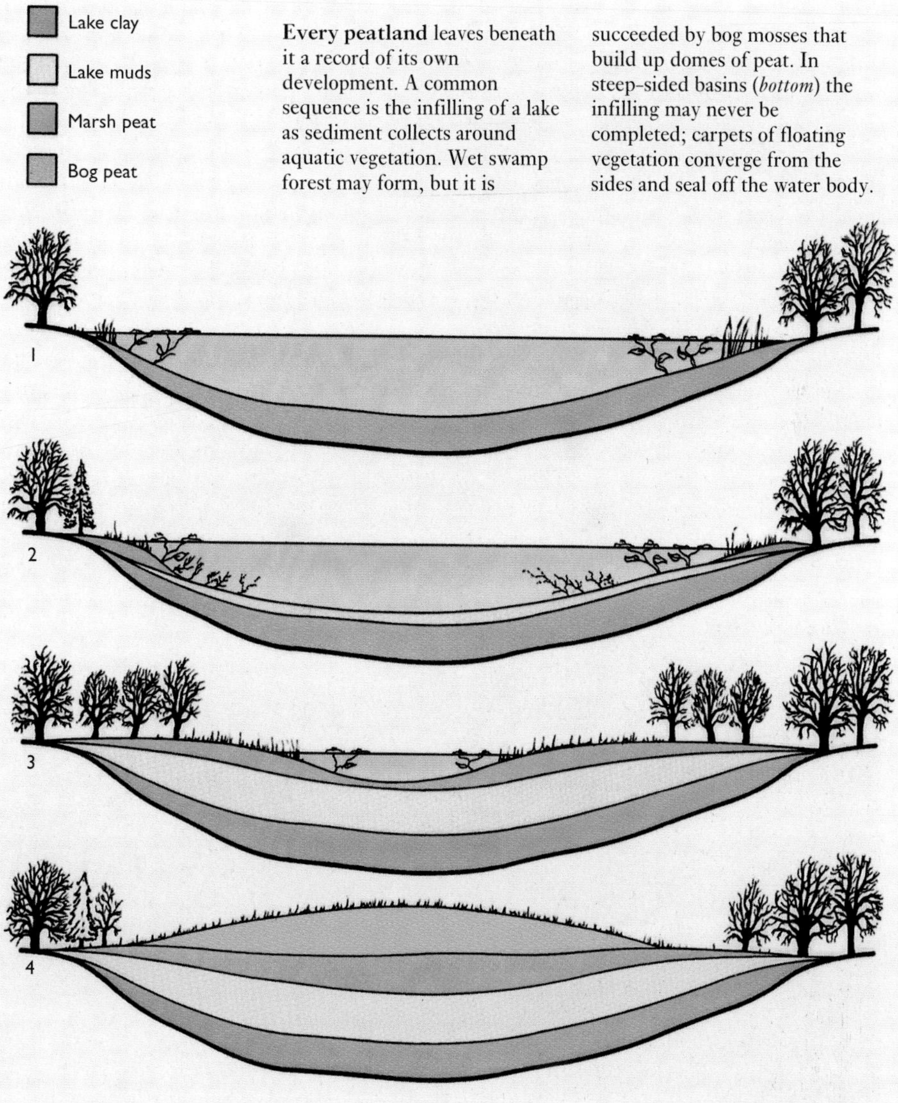

- Lake clay
- Lake muds
- Marsh peat
- Bog peat

Every peatland leaves beneath it a record of its own development. A common sequence is the infilling of a lake as sediment collects around aquatic vegetation. Wet swamp forest may form, but it is succeeded by bog mosses that build up domes of peat. In steep-sided basins (*bottom*) the infilling may never be completed; carpets of floating vegetation converge from the sides and seal off the water body.

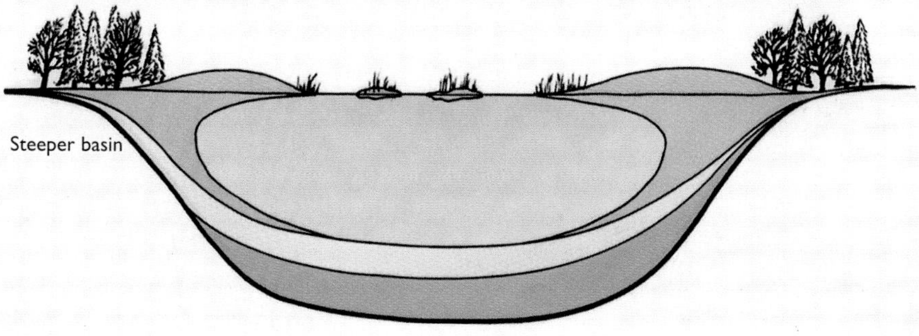

1

2

3

4

Steeper basin

Saltwater wetlands

The saltwater wetlands that fringe some coastal regions are biomes of constant development and change. These wetlands are found only along calm protected shores. Here, often at the mouths of estuaries where currents counterbalance to produce lazy eddies, fine particles of mud and sand settle out in mudflats and mudbanks.

These muddy environments form an anchorage for pioneer plants. The first plants to colonize estuarine mud need to be able to cope with a very unstable environment. The mud, which may move as water currents alter, may be covered with seawater for most of the day. When exposed to the air (normally twice a day), it may be baked by the sun or affected by frost. Exposure to air also results in fluctuating temperatures, and in increased salinity as the water evaporates. But if it rains at low tide, salinity decreases.

Eelgrass, glasswort, and cord grass are the most frequent colonists in temperate regions. Of these, cord grass is probably the most successful worldwide. This perennial can invade by seed and by fragments of rhizome (underground stem), and soon establishes itself. Although the glassworts are succulent, salt-tolerant, and capable of surviving both desiccation and immersions, these plants (mainly annuals) do not bind the mud as effectively as cord grass.

The presence of plants helps stabilize the mud and leads to its increased accumulation. As the mud surface gradually

The prop-rooted *Rhizophora* **tree dominates this Fijian mangrove swamp.**

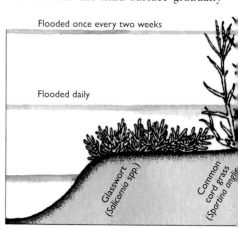

Flooded once every two weeks

Flooded daily

Glasswort
(Salicornia spp.)

Common
cord grass
(Spartina anglica)

rises, other plants arrive that are less tolerant of lengthy immersion, including saltmarsh grasses, arrow-grasses, and sea aster. The plants that eventually form a dense carpet of vegetation on the saltmarsh must all be tolerant of high salinity, and several are equipped with special glands that they use to get rid of salt.

The saltmarsh turf contains some insect-pollinated, nectar-producing plants such as thrift and sea lavender that attract insects to the community. Many other invertebrates find shelter in the turf, including small periwinkles and crustaceans. Where water drains from the marsh, soft mud is exposed in creeks or channels. Here wading birds such as dunlin, stilts, yellowlegs, and godwits feed. Small mammalian grazers, including hares, rabbits, and voles, visit the marsh and are preyed on by birds such as short-eared owls and kestrels.

Mangrove swamps

In tropical regions, coastal muds are often colonized in the first instance by mangroves, trees with leathery leaves. Their roots are adapted to the stagnant, airless muds they inhabit, and some grow upward into the air to collect oxygen. Prop or buttress roots provide stability where mud is mobile and high winds are frequent.

Mangrove seeds are adapted to the problems of pioneering—some begin to germinate before they fall, to insure quick establishment. To aid anchorage some seeds are spear-shaped and

stick fast in the mud where they drop.

Mangrove swamps are the saltwater equivalent of the tropical rain forest but do not contain the same diversity of plant life. The different mangrove trees do, however, form zonation patterns along the shore in response to such factors as mud stability, salinity, water depth, and local rainfall. In Malaysia, for example, the mangrove *Avicennia* usually colonizes saline mud first and occupies the outer zone of mangrove forest. Closer inshore is *Rhizophora*, which has longer prop roots but is less salt tolerant.

Fish and fish-eating birds such as egrets and anhingas abound in the mangroves. Saltwater crocodiles are the larger predators in swamps of Southeast Asia and northern Australia. Fish, such as the mudskippers, can drag themselves from the mud and clamber up protruding roots, and fiddler crabs scuttle over the exposed mud at low tide.

Habitats under threat

Among the greatest modern threats to saline wetlands is the development of coastal regions for shipping and industry, particularly since saline lands are usually of little agricultural value. Mangrove swamps, communities with a history of more than 60 million years, are even more at risk because they can be converted into rice paddies, as is happening in Southeast Asia. They also provide a source of fuel, and mangrove stripping is taking place in the Philippines, Malaysia, and Thailand.

Mangrove swamps around tropical coasts play an unexpected protective role. They absorb much of the shock of tidal waves caused by hurricanes and protect inland, populated areas.

When mangrove swamps are destroyed, they leave coasts vulnerable to damage by such storms. For this reason, mangroves are being planted at some sites in Bangladesh.

The plants that grow on this European saltmarsh are determined by the height of the soil surface. The highest areas are rarely flooded by the sea and support rushes and brackish water species. Low-lying sites are flooded with each high tide, and only plants that are tolerant of saltwater can survive.

on saltmarsh grass (*Puccinellia maritima*)

Creek

Sea purslane (*Halimione portulacoides*)

Sea plantain (*Plantago maritima*)

Thrift (*Armeria maritima*)

Sea lavender (*Limonium vulgare*)

Saline pan

Thrift (*Armeria maritima*)

Sea plantain (*Plantago maritima*)

Sea milkwort (*Glaux maritima*)

Red fescue (*Festuca rubra*)

Sea milkwort (*Glaux maritima*)

Saltmarsh rush (*Juncus gerardi*)

Oceans

Most of the water on Earth is in the oceans that dominate the planet. They cover 71 percent of the world's surface and, on average, are 12,700 ft or some 2.5 miles deep. These oceans provide a vast reservoir for the constant cycling of the Earth's water. When it evaporates from the ocean surface, water is transported in the air as water vapor, comes back to the ground as rain, snow, or dew, and is then returned to the seas via thousands of rivers.

The cycle takes the water through and past the rocks and soils of the landmasses. This steady percolation, which has been taking place for some four billion years, slowly dissolves soluble salts out of the Earth's crust and washes them into the sea. The saltiness of seawater results from the accumulation of these salts.

Seawater contains a huge variety of dissolved elements. Just six, however, make up more than 99 percent of the 1 oz of solid salts in any 2 lb of seawater. These are chlorine, sodium, sulfur (as sulfate), magnesium, calcium, and potassium.

Other elements are present at much lower concentrations, yet some are of vital importance to the creatures living in the seas. They provide mineral nutrients for the plant life of the oceans, particularly the phytoplankton organisms. These are the tiny plant cells that live and photosynthesize in the upper sunlit layer.

Microscopic phytoplankton are the main primary producers of the open oceans; there are virtually no visible plants except for the algae of the shoreline fringes. Like green plants on land, planktonic plants trap sunlight and turn dissolved carbon dioxide into structural and energy-providing organic molecules by photosynthesis. Further such molecules are constructed with the addition of other elements such as sulfur and phosphorus obtained from mineral salts. Almost all other life forms in the sea depend on this organic production, directly or indirectly.

Key requirements for life

All plants, on land and in the sea, depend on sunlight, water, carbon dioxide, and mineral salts to live. In the upper layers of the ocean, the first three of these needs are easily met, but mineral salts may be in short supply because they are used up by algae. Among the key elements on which productivity depends, but which may be limited, are phosphorus, in the form of dissolved phosphates, and nitrogen as nitrates.

Both phosphorus and nitrogen usually occur in the sea at concentrations of less than one part per million. They play central roles as plant nutrients and as such can determine the productivity (see pp. 56–57) of any part of the oceans. High or low levels of phosphates or nitrates make

the difference between "dead waters"—with little plant and animal life—and productive waters. In the latter, a sufficient supply of mineral salts supports a flourishing phytoplankton population that is fed on by zooplankton—microscopic marine animal life. This in turn provides food for larger sea animals such as small fish and squid which are themselves consumed by larger predators.

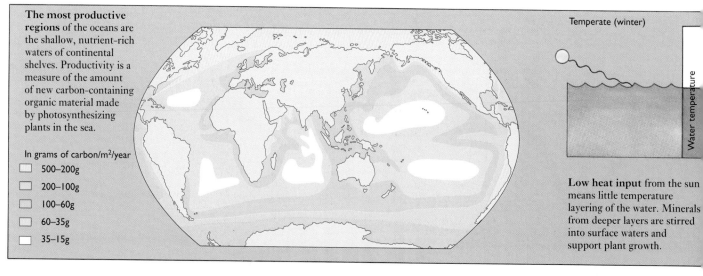

The most productive regions of the oceans are the shallow, nutrient-rich waters of continental shelves. Productivity is a measure of the amount of new carbon-containing organic material made by photosynthesizing plants in the sea.

In grams of carbon/m²/year
- [ ] 500–200g
- [ ] 200–100g
- [ ] 100–60g
- [ ] 60–35g
- [ ] 35–15g

Temperate (winter)

Water temperature

Low heat input from the sun means little temperature layering of the water. Minerals from deeper layers are stirred into surface waters and support plant growth.

The richest marine life occurs in shallow waters close to continents where plant productivity is high. Complex food webs carry the productivity of plant plankton via animal plankton, small invertebrates, and fishes to top predators such as sharks and tuna.

Shark

Tuna

Cod

Cuttlefish

Herring

Shrimp

Copepod

Animal plankton

Plant plankton

On land the areas of greatest plant productivity are in the tropics, but this is not so in the oceans; the areas of highest productivity are in cooler waters. This is because mineral salt replenishment is affected by the temperature and density of seawater.

In the tropics, for example, the sunwarmed surface waters encourage plant growth, and mineral levels are quickly depleted. The surface warming has the effect of making the upper layers of water less dense than the deeper, cooler water, away from the sun's rays. This lighter water, lower in salts, remains at the surface and, because of the density differential, cannot easily be replenished by the heavier, nutrient-containing water below. Consequently, tropical midocean zones are not highly productive.

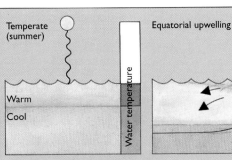

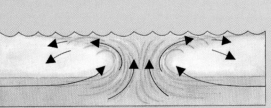

Temperate (summer)

Equatorial upwelling

Shore upwelling

Warm

Cool

Water temperature

In tropical and temperate summer seas the sun's heat produces layering of the water. The warm surface waters rapidly run out of nutrients and their productivity is low.

Near the equator different water movements to the north and south interact to push deep waters plus nutrients to the surface, thus improving productivity.

Where major water movements meet continents, deep nutrient-rich waters are brought to the surface. Such upwelling areas are often the site of rich fishing grounds.

The shore

The seashore is a highly diverse environment. And, although it may not always be obvious, it teems with life. The boundary between land and sea, it extends from the level of the lowest of low tides to the area above the highest tides, which may be reached by waves during the fiercest storms. The intertidal zone is the band of shore between these two tidal extremes. Above the high tide mark itself is the "splashzone" strip which may be affected by storm-driven waves and spray.

Tides are produced by the gravitational pull of the moon and sun on the oceans. These pulls produce bulges of water, approximately under the moon and roughly at the mirror position on the opposite side of the globe. In the course of the Earth's 24-hour spin around its polar axis, most points on the globe experience these two tidal bulges (high tides): most coasts have two high tides and two low tides in about 24 hours.

A central point on a shore, between high and low tide marks, is exposed to the air for two periods in a day and immersed in water for two periods. Either side of that point, the total time immersed varies from about 100 percent of the day near the low water mark to almost zero around the high tide level. Thus physical conditions on the shore are much more variable within a small area than in most environments. Just a few feet can mean the difference between an entirely terrestrial habitat and one that is intermittently marine.

Shore zones experience differing conditions—temperature, wave action, and most important, periods of immersion and exposure to air. Each has its communities of plants and animals adapted for particular levels of seawater immersion.

On rocky shores, where there are many places for the larger seaweed (algae) species to attach themselves, zones are clearly demarcated by these plants. A common sequence found in temperate climates begins with dark encrusting lichens on the splashzone rocks and green algae in the upper intertidal zone, where there may be some freshwater runoff. Below this is a zone dominated by brown algae, and at the low tide level and just below it, there are large brown algae and red algae. Other red algal forms live mixed among other types at all levels of the intertidal zone.

The attached algae of a rocky shore are the principal primary producers. They also provide physical habitats for other organisms, and a zoned pattern of attached and wandering animals often emerges. Some stick to the weeds themselves, others to the rocks; some enter rock crevices while yet others move over the available plant and rock surfaces. Characteristic attached animals are barnacles, sea

Rocky shores harbor a rich variety of plants and animals, each adapted for conditions in a particular zone. Those near the low tide mark spend much of their lives immersed in seawater; those around the high tide mark are usually exposed to the air.

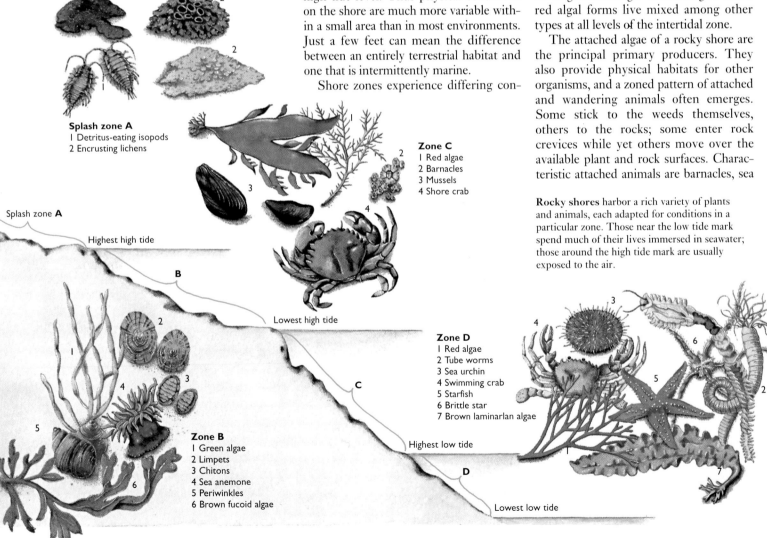

Splash zone A
1 Detritus-eating isopods
2 Encrusting lichens

Zone C
1 Red algae
2 Barnacles
3 Mussels
4 Shore crab

Splash zone **A**

Highest high tide

B

Lowest high tide

Zone D
1 Red algae
2 Tube worms
3 Sea urchin
4 Swimming crab
5 Starfish
6 Brittle star
7 Brown laminarlan algae

C

Highest low tide

D

Zone B
1 Green algae
2 Limpets
3 Chitons
4 Sea anemone
5 Periwinkles
6 Brown fucoid algae

Lowest low tide

anemones, hydroids, and shellfish such as mussels. The many mobile species include worms, mollusks, crustaceans, starfish, sea urchins, sea cucumbers, and brittle stars. When the tide is in, shallow-water fishes move in among the weeds to feed on other intertidal animals.

Sandy shores

Shores without rocks do not appear to be as packed with life. Whether composed of sand, mud, or a mixture of the two, they usually lack large surface seaweeds because there are few firm anchorage points in the shifting sediments. Zonation is therefore less evident. The main primary producers of a sand or mud shore are microscopic organisms such as the yellow-brown diatoms. These occur in the water itself or on the sediment surface.

The animals of sediment shores are usually burrowers such as worms, bivalves, sand dollars, brittle stars, sandstars, and others. Almost every marine animal group has evolved specialists that can construct tubes or dig tunnels in marine bottom sediments. In this way they protect themselves from the worst effects of exposure at low tide.

On sandy and muddy shores there is far less solid substrate for animals such as mollusks to attach themselves to. Most live underground, burrowing into the sediment.

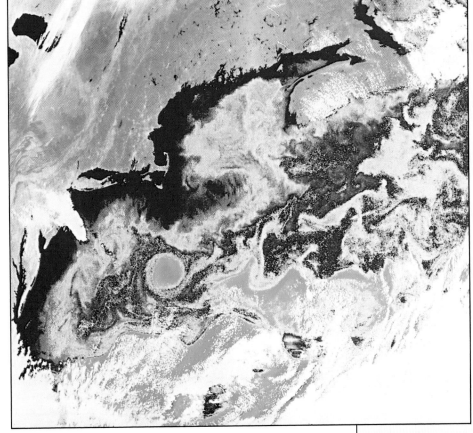

A false-color satellite image of the northwest Atlantic shows areas of low productivity in blue. Green, yellow, red, and dark red indicate increasing levels of productivity toward land (buff).

Zone B
1 Scallop
2 Cockle
3 Furrow shells
4 Razor clams
5 Spiny cockle
6 Burrowing worm
7 Lugworm

Zone C
1 Sand dollars
2 Brittle stars
3 Heart urchins
4 Eelgrass

Lowest low tide

Highest high tide

A

B

C

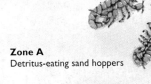
Zone A
Detritus-eating sand hoppers

The coral reef

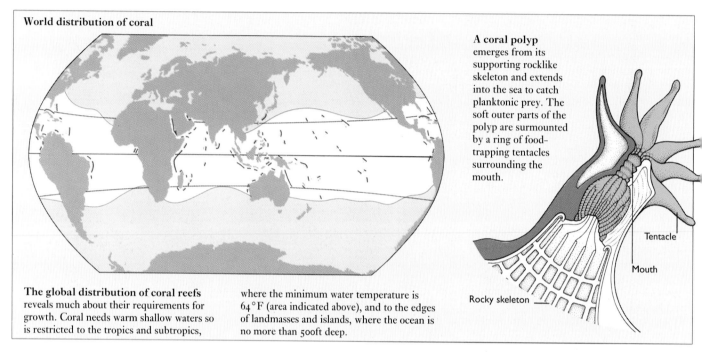

World distribution of coral

A coral polyp emerges from its supporting rocklike skeleton and extends into the sea to catch planktonic prey. The soft outer parts of the polyp are surmounted by a ring of food-trapping tentacles surrounding the mouth.

Tentacle

Mouth

Rocky skeleton

The global distribution of coral reefs reveals much about their requirements for growth. Coral needs warm shallow waters so is restricted to the tropics and subtropics, where the minimum water temperature is 64°F (area indicated above), and to the edges of landmasses and islands, where the ocean is no more than 500ft deep.

Like rock buttresses against the action of the sea, coral reefs can stretch more than 1,250 miles along tropical coastlines and support a fantastic variety of marine life. These massive reefs are built by other living things as well as the coral animals themselves.

The solid substance of a reef consists of the limestonelike (calcium carbonate) outer skeletons of coral polyps—invertebrate animals closely related to sea anemones. A normal sea anemone in a rock pool has a soft unprotected body, capped with a mouth surrounded by feeding tentacles. The soft multiheaded body of a coral polyp, however, is encased in a rocky protection, a skeleton which it secretes itself in the same way as a snail builds a shell. Each outcrop of coral has a base of the accumulated rocky skeletons of dead corals, with a surface coating of living coral polyps. These emerge through slits or holes in their skeletons to catch their food with their tentacles.

But coral animals alone cannot build the reef. They depend on their symbiotic (mutually dependent) partnership with tiny single-celled photosynthetic creatures that are also found free in the seawater.

The animals take the photosynthesizing organisms into their bodies, either in or between their own cells. These organisms gain a protected habitat in which to live and some nutritional mineral salts from the animals' body fluids. In sunlight, the animals are able to gain organic food which is produced by the photosynthesis of their guests.

The partners of the corals belong to a group of single-celled organisms called the dinoflagellates. Inside corals these cells do not lose their power to photosynthesize but continue to trap sunlight energy and synthesize organic food molecules such as sugars. Some of these food molecules pass into the coral and help its growth.

The most crucial effect of the dinoflagellates is to enable the corals to convert calcium salts in seawater into a rocklike calcium carbonate structure. Without their partners, corals would be just soft colonial sea anemones, and reefs would not exist.

Corals have certain habitat requirements that constrain the geographic distribution of coral reefs. As invertebrates adapted to tropical conditions, they require warm seawater, and, because their

feeding tentacles can become clogged by any sediment, the water must be clean and clear for maximal growth. Corals also need a hard sea bottom on which to attach their heavy skeletons and sunlight for their photosynthesizing partners. They never occur at depths of more than 500 ft; below this there is not enough light.

There are three main types of reef: fringing reefs, barrier reefs, and atolls. Fringing reefs are found close to the edges of a landmass in shallow waters. Barrier reefs are more extensive structures that are separated by a lagoon from the continent or island landmass. Atolls are approximately circular reefs that appear to have formed around volcanic islands that subsequently sank.

The complex creviced and branching shapes of coral outcrops provide an environment that offers many opportunities for colonization for huge numbers of other plants and animals. Mollusks, crustaceans, and many other invertebrates find homes there, and a rich variety of colorful fishes weave their way through the coral world. As a substrate for other life forms, the corals of the reef play the same role as the trees of a forest.

A reef community off the coast of Papua New Guinea.

Niche patterns

Worlds within worlds

Any one habitat provides a multitude of different opportunities for an animal to earn a living. Take, for example, a coral reef. In environmental terms, it is a relative uniform place. Its margins are precisely defined, cut off at t top where the pounding of the waves at low tide breaks the coral branches, and at the bottom by the depth to which sunlight can penetrate, for coral cannot live without light. Between these two limits, the temperature of the water, its saltiness, its clarity, and the amount of oxygen dissolved in vary only within quite narrow limits. Yet even if you overlo the multitudes of crabs, sponges, starfish, worms, mollusks, much else besides, and consider only the fish of the reef, th variety is enormous.

Each has its own particular diet. Some graze the marine algae, some sieve tiny floating organisms from the water, an still others hunt smaller fish. Parrot fish crunch up the cora itself, pulverizing the stony skeleton in order to extract the edible bodies of the minute coral polyps—and they have powerful rounded teeth to enable them to do so. Others hav jaws elongated into long slender forceps so that they can delicately pick off little worms and other morsels from the nooks and crannies of the coral. Some even graze the skins other fish and live by eating dead skin and parasites.

Nor is their specialization limited to their diets. It extend their resting and hiding places. Little blue damsel fish favor branches of a particular species of coral and dive to safety between its sharp stony arms whenever danger threatens. Th are seldom found anywhere else except close beside it. Trigg fish use narrow clefts as bolt holes and lock themselves insid by erecting their trigger, a spine on their back, so that they securely jammed between the two sides. Gobies dig holes, moray eels live in caves, and clown fish consort with sea anemones, wallowing in their tentacles, protected from pred by the anemone's poison stings, to which the fish themselve immune. One remarkable fish actually lives within the gut o sea cucumber and moves in and out of its lodgings as it plea by using a special touch, as though ringing a door bell, to stimulate the sea cucumber to relax the muscles around its v and allow the fish to pass through.

Such variables as there are in the physical environment of reef give fish more chances to specialize. Some fish are activ only during the day, others only at night. Some prefer the o side of the reef, where the water is slightly cooler and more oxygenated. Others live on the landward face, where there is less oxygen and warmer water, and they will die if they are transported into what might seem to be better conditions.

This assumption of particular professions by different species is found not only in coral reefs but in every inhabited environment that exists. This is so whether the community it supports is highly complex, such as the tropical rain forest, or whether it is so harsh and impoverished that only a few species manage to survive in it, as is the case in the high Arctic, where there is darkness for much of the year.

Any one species of animal, therefore, possesses a set of specializations that determine how and where it spends its life—its diet, its home, its physiological preferences of temperature, and its patterns of activity. Taken together, such factors define its niche. In this competitive world, no two species can occupy exactly the same niche for long. One would inevitably be marginally more efficient and would therefore ultimately displace the other. Should it appear to us that two different species living alongside each other have identical lifestyles, close study will almost certainly reveal that they are exploiting their environment in slightly different ways.

Ants and anteaters

But in different parts of the world, quite unrelated species may occupy equivalent niches in the community. Ants and termites are one of the most abundant sources of protein to be found anywhere in the tropics. The problem is how to collect them. Most live in fortresses of one kind or another, either in underground galleries or in concrete-hard mounds of their own construction. One anatomical device is more efficient than any other at extracting them—an extremely long muscular tongue covered in glue. That is exactly the equipment used by the giant anteater of South America. Its jaws are elongated into a curving tube and its tongue can extend for 2 ft beyond its mouth. The animal wanders across the grassy savannas at night, ripping open the ants' nests with its immensely powerful forelegs and flicking its tongue deep along the galleries of the nests.

In Africa this niche of nocturnal ant and termite feeder is occupied by a quite unrelated creature, the aardvark. Yet it, too, has evolved just such a tongue. In Australia, a third mammal has specialized in feeding on ants and termites. It is a marsupial more closely related to the kangaroo than to either the giant anteater or the aardvark. But even though, compared to them, it is very small, being no bigger than a rabbit, it has developed that long narrow snout and that long sticky tongue. So the occupation of a similar niche tends to result in animals in different parts of the world developing similar physical characteristics.

Their similarities may, indeed, extend beyond their appearance to such seemingly unconnected characteristics as the way they reproduce. In all tropical rain forests, occasional trees rise high above the general level of the canopy. They stand out like islands in a sea of leaves. In them you are likely to find a gigantic nest and standing on it a huge eagle. It lives by hunting across the surface of the canopy, catching any monkeys that are incautious enough to climb in the topmost branches. Its wings are broad and relatively short and it has a long tail, features that give it the great maneuverability it needs when diving between the branches. It has huge claws with which to seize its prey and a massive beak with which to butcher it. The bird is immense, as it has to be to have the strength to carry off such big victims. You can see such birds in the forests of South America—the harpy eagle; in Southeast Asia—the monkey-eating eagle; in Africa—the crowned eagle. Take one from each of these forests and put the three birds side by side in a zoo and you would have to be very expert to tell one from the others, even though they belong to different genera in the eagle family.

But occupying the same niche has also given them similar nesting habits. Their young must grow to a considerable size before they are able to hunt for themselves. That takes a long time and requires the parent to catch and bring back to the nest great quantities of meat. So all three species rear only a single youngster at a time and all have to feed it for almost a year before it leaves the nest.

Exploiting niches

The more food an environment produces, the more inhabitants it can support and the more crowded it becomes. And the greater the crowd and therefore the competition, the more specialized animals tend to become, and the narrower the niche they occupy. No habitat seems so remote, no food so indigestible, no conditions so uncomfortable but that some animal has developed specialisms that enable it to exploit them. That seems a very extravagant statement, yet again and again naturalists come across a species that has taken up a way of life that seems beyond the wildest flights of human fancy. Consider this: in the densely crowded environment of tropical Africa, a worm flourishes that lives only under the eyelids of hippopotami and feeds only on their tears. That really is a narrow niche.

The story of
the crossbill

Norway spruce (*Picea abies*)

☐ Red crossbill

The red or common crossbill (*left*) is widespread in both the Old and New Worlds. Its distinctive bill is ideally suited to the extraction of seeds from cones, particularly spruce.

Every organism has its niche—its role in the community it occupies. A description of an animal's niche would include where it lives, what it eats and how it obtains that food, what its environmental requirements are, and so on. Some niches are broad and include many variables while others are much more specialized.

The crossbill's niche
The red or common crossbill (*Loxia curvirostra*) is a bird with a narrow, specialized niche. This small gregarious finch has a wide distribution and is found throughout the northern boreal forests. It also lives in coniferous woodlands in mountain areas such as the Alps and the Rockies. The crossbill is a bird of the forest canopy and is rarely seen on the ground except when it lands at a pool to drink.

The crossbill's most distinctive feature is its beak—the upper and lower mandibles are crossed at their tips. This shape has evolved in response to the bird's feeding habits: it pries seeds out of the cones of trees such as spruce, larch, and pine, using its crossed bill to force open the scales of the cone while it extracts the seed held between them with its tongue.

Although this is a highly specialized diet, cones and their seeds are widely available, so crossbills are both wide-spread and common. But the crossbill can take few other foods. It will eat berries, buds, thistle seeds, and even insects, but it is far less efficient at manipulating these. It is hard, therefore, for the crossbill to colonize new habitats, and it is largely restricted to coniferous forests and a diet of cone seeds.

Feeding adaptations
Even within such a narrow niche there is room for further specialization. Other species and races of crossbill have evolved with bills adapted to their specific feeding needs. Over much of Europe common crossbills feed on the seeds of spruce cones. But in some areas, such as Scotland,

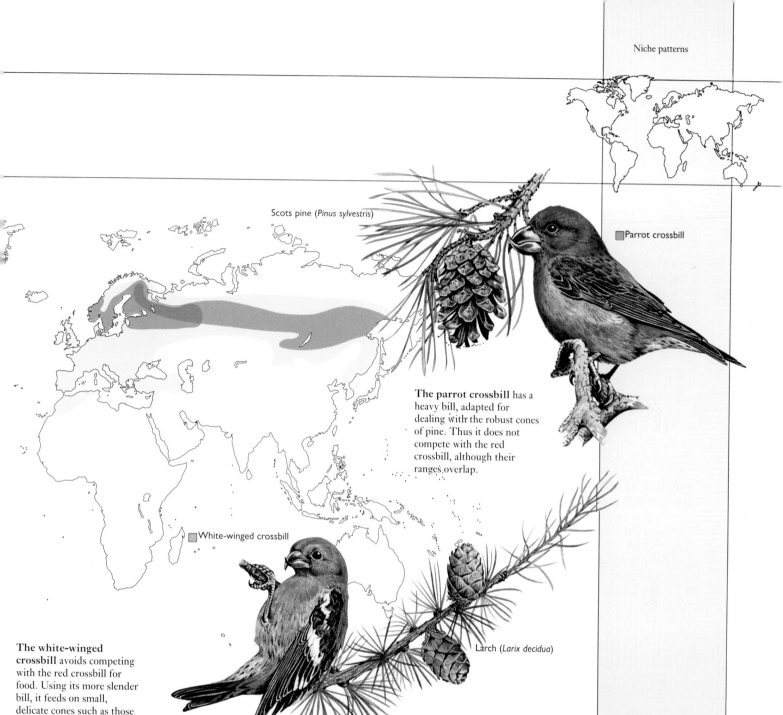

Scots pine (*Pinus sylvestris*)

Parrot crossbill

The parrot crossbill has a heavy bill, adapted for dealing with the robust cones of pine. Thus it does not compete with the red crossbill, although their ranges overlap.

White-winged crossbill

Larch (*Larix decidua*)

The white-winged crossbill avoids competing with the red crossbill for food. Using its more slender bill, it feeds on small, delicate cones such as those of larch.

some Mediterranean islands, and North Africa, other types feed on cones with tougher scales such as those of the Scots pine and the umbrella or Swiss pine. These birds have evolved thicker, heavier bills to deal with their food. The heaviest bill of all belongs to the parrot crossbill (*Loxia pytyopsittacus*), which feeds on Scots pine cones in Scandinavia and northern Asia.

Thin-billed types have become specialized to deal with more delicate cones. The white-winged crossbill (*Loxia leucoptera*) of the Asian forests, for example, feeds on larch cones. Thus it has evolved to fit a subtly different niche from its thicker-billed relatives.

In North America there are just two crossbill species, the red crossbill and the white-winged crossbill, with its thinner bill. Resources are shared between them, the red crossbill taking the heavier cones, usually pine, while the white-winged prefers the lighter cones of spruce and tamarak.

The crossbill, then, is very successful within an extremely narrow niche. Its bill structure has become so highly adapted that it is ill-equipped to deal with any food apart from cones. And its diet restricts it to the coniferous forests where its food is plentiful. Although the cone is a tricky fruit to exploit, it is such an abundant food resource that it is worth the evolutionary effort and the risks involved.

The expansion of coniferous forestry has meant that the crossbill is spreading into previously uncolonized areas of Europe and North America. If suitable habitats and food sources are found, birds may establish new colonies and thus extend their ranges.

In the United States the red crossbill is currently spreading southward from the boreal zone on the East Coast and breeds as far south as Georgia.

Day shift/ night shift

Competing species in an environment often "divide up" the existing resources in order to maximize their use. Several bird species may coexist in the same area of forest but find their food at different levels or in different parts of the trees.

Less obvious as a resource than food or nesting sites is time. The clearest and most common example of "time-sharing" is that between daylight-active (diurnal) and nighttime-active (nocturnal) animals in the same environment. Both groups can live in the same space, searching for exactly the same food, but one comes out at night while the other sleeps.

A rhythm of light and darkness

Almost everywhere on the Earth's surface there is a 24-hour rhythm of light and dark. The main exceptions include the deeper regions of the oceans, where no light ever penetrates, and the deepest caves.

The durations of dark and light periods in any day are variable. They change with the seasons and they change with latitude. Close to the equator, day and night lengths are about 12 hours each and relatively unaffected by the seasons. Nearer the poles, because of our planet's tilted spinning axis, day and night durations are much more variable with the time of year. But no matter what their lengths, these days and nights are the backdrop against which diurnal and nocturnal animals play their respective scenes.

There are day-specialist and night-specialist animals in almost all ecosystems. Hawks hunt rodents and other small creatures during the day; owls take over at night. Most butterflies are in the air by day, to be replaced by moths at night.

But how do these specialists evolve? For any animals with vision, which includes most invertebrate and vertebrate species, lit and unlit environments pose quite distinct problems and opportunities. Nighttime darkness is an enormous disadvantage for a creature that senses the presence of its prey or its predators by sight, unless it has additional sensory abilities. This sensory imperative has led

The vervet monkey (*Cercopithecus aethiops*), a primate largely active by day, shares its habitat with the nocturnal bushbaby. It finds its food in the trees but also forages on the ground. Although its diet is mostly plant-based, it eats invertebrates, lizards, birds' eggs, and nestlings.

The greater bushbaby (*Otolemur crassicaudatus*) is a specialist nocturnal primate of African forests and wooded savannas. It sleeps in the trees by day and emerges at night to hunt for spiders, insects, and young birds and to feed on plants.

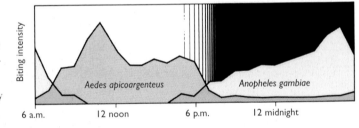

Mosquitoes bite by day or night, depending on their species, to obtain their meals of human blood. The graphs show the biting cycles of two common species. They only overlap briefly at dawn and dusk.

Biting intensity

Aedes apicoargenteus — *Anopheles gambiae*

6 a.m. — 12 noon — 6 p.m. — 12 midnight

to the evolution of specifically nocturnal animals with high sensory abilities in the dark. Owls, for example, have huge ultra-sensitive eyes that are responsive to even the lowest light levels. They also have extremely acute hearing and can pinpoint the sources of minute sounds.

On the other hand, the problems of nighttime activity have encouraged certain patterns of behavior in diurnal animals; they sleep at night to maximize their safety during the hours of darkness and their efficiency during the day.

The behavior patterns of day- and night-active animals can be triggered by the light changes at dawn and dusk. They are switched on by rapidly changing light levels. Alternatively their behavior may be controlled by internal biological clocks, probably present in the brains of most animals, which tick with a roughly 24-hour

rhythm. These can produce a change of behavior for night or day even without the trigger of external light clues.

The nighttime specialists

The bats are an example of a major group of animals that has apparently arisen and diversified successfully as a result of adaptations linked with night specialization. About one mammal in four in the world is a bat, and bats are the only mammals capable of sustained flight. This is accomplished by means of efficient wings that are constructed from flight membrane stretched between the elongate fingers of the hands and the sides of the body.

The bats' success as a group seems to be linked to their capacity to fly and catch flying prey at night, even in complete darkness. No other vertebrate can do this with anything like their facility. As a

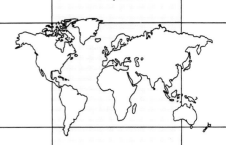

Some blood-transmitted parasites have become adapted to the specific nighttime biting activity of certain mosquitoes. The nematode worm *Wucheria*, which causes the disease elephantiasis, appears in human blood only at night—when it can be picked up by the night-biting mosquitoes that transmit it from person to person.

result, they share the skies with the other predators of flying insects—the birds. This sharing is generally done on a strict, two-shift basis. The birds use the air by day, the bats by night. Only a few bird species, such as owls and nightjars, provide a little nighttime competition.

The supreme nocturnal capabilities of bats are based on a unique sensory system that involves flight guidance and the location of flying prey in darkness by a form of "biological radar." Bats make ultrasonic squeaks and churrs at pitch levels way above the hearing range of the human ear. These sounds bounce off objects and the returning echoes are picked up by the large, aerial-like ears of the bat, which can detect ultrasonic frequencies. The returning sound pulses are analyzed to give the creature a sound echo map of its surroundings to fly by; they also act as a homing radar to track flying prey.

This sensory "technology" was an evolutionary breakthrough for the ancestors of today's bats. It meant that they had the nighttime sky and the insects it contained to themselves. An enormous and previously almost empty niche space was opened up for them and they adapted to fit it. So phenomenal has been their success that there are now approaching a thousand species of these nighttime specialists.

Bats and other nocturnal creatures are an example of the slow yet constant adaptation of living things to the world around them as they jostle for position in crowded ecological space. Any way of life or resource that is not fully exploited constitutes a gap in that space. Steady evolutionary change can produce new, specialized organisms that can insert themselves into these unused niches.

The greater horseshoe bat (*Rhinolophus ferrumequinum*) has a wingspan of up to 14 in and is an expert nocturnal hunter. As well as feeding on airborne insects, these bats are adept at using their ultrasonic radar to home in on beetles on the ground. They fly low, locate their prey, and swoop on them with extraordinary accuracy.

Competition and survival

Studying the behavior and capacities of an animal species in the laboratory under strictly controlled conditions provides a useful but limited view of the way that animal lives. For this type of study can produce little more than a list of observed behavior and physiological abilities. It is a list that rarely catches the true essence of the animal's way of life because it is based on a species in isolation; in the real world, species live in rich, complicated communities of animals, plants, and microbes.

It is only possible to understand an animal's actual lifestyle, or niche, if it is investigated within the community of other species among which it exists. This is the living environment in which its behavior and physiology have evolved.

When the niches of the species in a community are described—whether of rodent species in a field, canopy birds in a wood, or carrion scavengers in savanna country—a remarkable patterning of niche types nearly always emerges. The niches of the coexisting animals do not "overlap" much. That is to say, when the community of animals is looked at as a whole, habitat resources seem to be neatly apportioned between the different species. The division is never absolute, but overall it does seem as though the niches of even closely related species in the community are significantly different from one another.

Dividing resources

Reference to a specific example will make this patterning clearer. In the 1960s, two ecologists made careful studies on the island of Trinidad of the niches of eight coexisting species of tanager—brightly colored songbirds of the New World tropics. Of the eight species, three, the speckled (*Tangara guttata*), the bay-headed (*T. gyrola*), and the turquoise tanager (*T. mexicana*), were extremely closely related. They all belonged to the same genus, lived in the same trees and bushes, and fed on insects and fruit. This suggests little in the way of a division of resources, for all three species seemed to

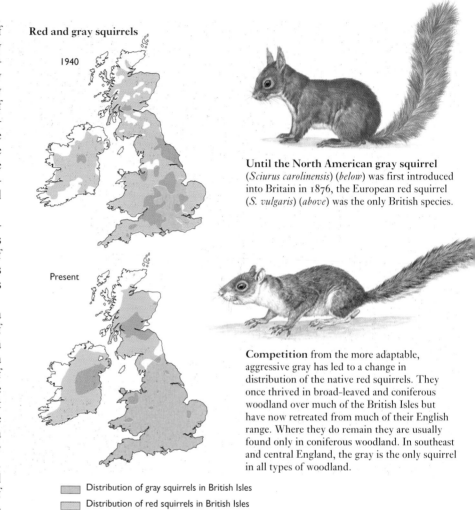

Red and gray squirrels

1940

Present

Distribution of gray squirrels in British Isles
Distribution of red squirrels in British Isles

Until the North American gray squirrel (*Sciurus carolinensis*) (*below*) was first introduced into Britain in 1876, the European red squirrel (*S. vulgaris*) (*above*) was the only British species.

Competition from the more adaptable, aggressive gray has led to a change in distribution of the native red squirrels. They once thrived in broad-leaved and coniferous woodland over much of the British Isles but have now retreated from much of their English range. Where they do remain they are usually found only in coniferous woodland. In southeast and central England, the gray is the only squirrel in all types of woodland.

be using the same ones. More detailed field observations, though, showed up the niche differences, as is clearly demonstrated by considering one aspect of the pattern of resource division.

In hunting for small insect prey in vegetation, the speckled tanager almost exclusively searches the leaves themselves. It clings to them upside down, picking off insects, or it walks along small twigs, picking off insects from the leaves above it. The other two species only rarely feed like this. Both obtain most of their insect prey from the undersides of branches. The bay-headed species does this mainly on quite substantial branches, hopping along and leaning over each side alternately to reach under it for insects.

The turquoise tanager, in contrast, almost always takes insects from fine twigs, usually those less than half an inch in diameter. It also has a predilection for the insects found on dead twigs, which are usually untouched by the other two species. These detailed observations show that insect food resources and specific feeding areas on the island of Trinidad are neatly split even between very closely related birds.

Findings such as these have led to the expression of what has become almost an article of dogma about niches and communities: if two animals have the same niche characteristics, they cannot coexist for long in a stable way in the same habitat. This is because, in a direct fashion,

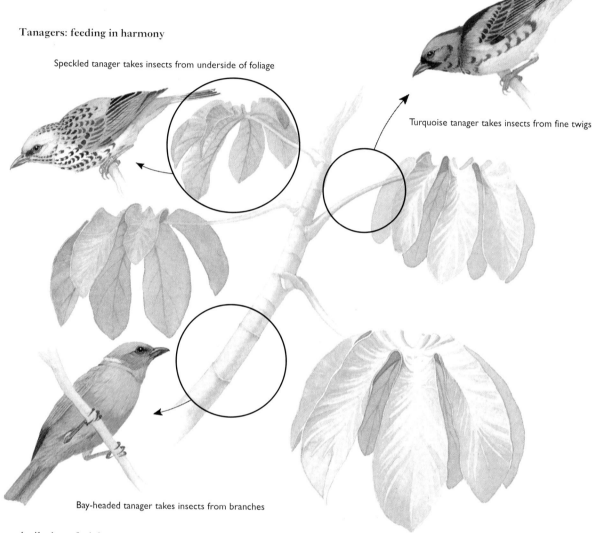

Tanagers: feeding in harmony

Speckled tanager takes insects from underside of foliage

Turquoise tanager takes insects from fine twigs

Bay-headed tanager takes insects from branches

Mowing a lawn may actually increase the incidence of weeds. If they are constantly cut back, all the plant species in the lawn—both weeds and grasses—are stimulated to grow again. If the lawn is left alone, the grasses will eventually dominate and suppress most other species. But, in the continued absence of mowing, woody shrubs and trees will become established and will eliminate the grasses by shading them.

similarity of niches means competition for habitat resources.

And great niche similarity means immense competition. In a condition of all out "war" between two species with very similar niches, it is almost inevitable that one will be slightly more efficient in some way. This means that it will breed more successfully and supplant the less efficient species.

The lifestyle of an animal that exists without any real competition in its environment is sometimes called its "fundamental niche." It can be thought of as the raw niche material that can be modified when the species finds itself in a community of competing species. When this transition in circumstances occurs, it is

common for the niche of at least one of the competing species to "contract," with the result that it uses only a part of the range of resources it has previously used.

Competing squirrels
A good example of this is the contraction in woodland range of the native red squirrel in the British Isles after the introduction of the American gray squirrel. The indigenous red squirrels have been forced to retreat, largely because of direct competition for food and breeding territories. In the southern half of mainland Britain, they are now restricted to fringing pockets of mainly coniferous woodland and in some areas they have disappeared altogether.

These three species of tanager coexist in the same woodland on the island of Trinidad. Although they all feed on insects in trees and shrubs, they do so in subtly different ways. This means they do not use quite the same food resources. Hence competition is reduced and the birds have a better chance of survival.

Adapting to surroundings

It is common to find that animals living together in one habitat occupy niches that differ from one another in clear but often subtle ways. They appear to be avoiding excessive competition for resources such as foods, space, and nesting territories.

There are also other types of niche patterning. One particularly different type emerges when some important physical characteristic of the landscape—such as temperature, wave exposure, water availability, oxygen, or light—varies in a predictable spatial way across the habitat. This physical change can be envisaged as a gradient, with high values of the character in one part, tapering to low values in another.

Lifestyle strategies

Such environmental gradients, via the mechanism of evolutionary change, have induced a response in the niche patterns of the animal species that live among them. Along the length of the gradient, it is common to find a regularly spaced series of related species, each of which has become specialized for maximum efficiency in one specific segment of the gradient and lives mainly in that geographical area.

A lifestyle strategy of this sort is, however, only valuable when the gradient exists, more or less unaltered, for the lifetime of many generations of the animals concerned. In randomly changing conditions, it would be dangerous for an animal species to specialize. If it did, although it might still operate highly efficiently for part of the time, when conditions changed it could face local extinction.

The scale of the gradient and the sequence of the species that have matched themselves with sections of it are intriguingly diverse. They may be on a tiny scale: for instance, along the gut of a single animal there are predictable gradients of many important variables that are crucial for parasitic worms living in that animal's gut. Acidity, for example, is often high in the stomach, but gradually declines to alkaline pH levels lower down the intestine, and the parasites often become specialized for survival in just one of the microhabitats offered by the different zones.

Parasitological research work on the gut of the flounder, a flatfish, has revealed the complexity of spatial parasite patterning on the minutest scale. Different species of flukes (digeneans), roundworms (nematodes), and spiny-headed worms (acanthocephalans) are typically found in restricted portions of the flounder's alimentary tract: a gradient less than 3 ft long.

Adapting to cold

At the other end of the geographical scale range, it is possible to chart smooth changes in species adaptation that appear to be responses to a habitat gradient across an entire continent. Some of these types of changes occur so predictably that they have been termed biological "rules." One such is Allen's Rule, which states that as habitats become increasingly cold, so the extremities of warm-blooded animal species become correspondingly shorter and more compact.

This rule is clearly demonstrated by the hare and rabbit species found in a south to north sequence, moving from Central into North America. Antelope jackrabbits, black-tailed jackrabbits, snowshoe and arctic hares are distributed in zones that scarcely overlap, and the sequence shows a gradual and adaptive reduction in the proportional length of the ears and legs.

Midway between gut- and continent-sized gradients and species sequences are those on a geographical scale of tens or hundreds of miles. A good example of this is the pattern found along river estuaries, where many animal groups show a series of specialist species, each adapted to the main environmental gradient of varying salinity.

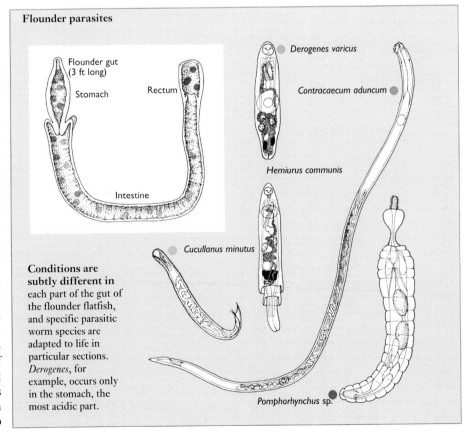

Flounder parasites

Flounder gut (3 ft long)

Stomach

Rectum

Intestine

Derogenes varicus

Contracaecum aduncum

Hemiurus communis

Cucullanus minutus

Pomphorhynchus sp.

Conditions are subtly different in each part of the gut of the flounder flatfish, and specific parasitic worm species are adapted to life in particular sections. *Derogenes*, for example, occurs only in the stomach, the most acidic part.

North American jackrabbits and hares

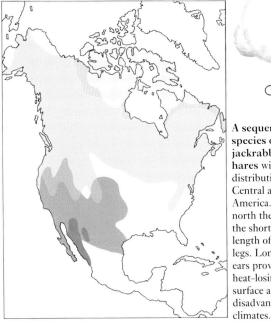

○ Arctic hare (*Lepus arcticus*)

A sequence of species of jackrabbits and hares with distinct distributions occurs in Central and North America. The father north the species lives, the shorter the relative length of its ears and legs. Long limbs and ears provide more heat-losing body surface and so are a disadvantage in cold climates.

○ Snowshoe hare (*Lepus americanus*)

◐ Black-tailed jackrabbit (*Lepus californicus*)

◐ Antelope jackrabbit (*Lepus alleni*)

Freshwater shrimp

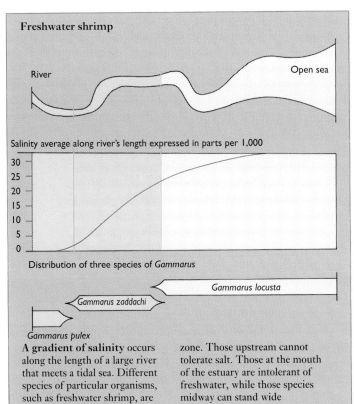

River

Open sea

Salinity average along river's length expressed in parts per 1,000

30
25
20
15
10
5
0

Distribution of three species of *Gammarus*

Gammarus locusta

Gammarus zaddachi

Gammarus pulex

A gradient of salinity occurs along the length of a large river that meets a tidal sea. Different species of particular organisms, such as freshwater shrimp, are adapted for conditions in each zone. Those upstream cannot tolerate salt. Those at the mouth of the estuary are intolerant of freshwater, while those species midway can stand wide variations in salinity.

Parallels in the bird world

Within a single habitat, one particular ecosystem, the differences between the adaptations of individual animal species are the most striking feature. These differences define the bounds of the niches of those species; they constrain competition between them.

Another perspective emerges, however, when habitats of similar type in widely separated parts of the planet are compared. Then, almost invariably, it is not the differences that are most remarkable, but the similarities of adaptive design in animals cast in similar roles in the ecosystem.

This is a fascinating example of a general phenomenon known as convergence or convergent evolution, which can determine any aspect of an animal's makeup. Two animals that share a pattern of function are likely to have evolved structural and behavioral similarities, even if they are not closely related. Indeed, such similarities may occur in completely unrelated animal types.

It is as though there is only a limited number of design plans that can fit any particular niche specification efficiently. Over and over again in animal evolution, the same solution to a particular design problem is turned up by the process of natural selection.

The answer to the problem "design a pair of jaws that can catch and hold a slippery, fast-moving fish prey" has more than once been long, tapering jaws set with many evenly spaced, pointed peglike teeth. Evolution, moving along many tracks, has "converged" on this answer.

Not only do some predatory fish, such as the garpike, have jaws like this, they also occur in almost any other vertebrate group, living or extinct, that has members feeding in this way. Fish-eating crocodiles, the gavials, have them, as did the now extinct swimming reptiles, the ichthyosaurs and the plesiosaurs. Diving, fish-catching ducks, such as the mergansers, and smew, are called "sawbills" because of their long, serrate-edged beaks. All exhibit the same design solution to a common problem.

The birds provide many examples of this remarkable pattern in animal life, and often the convergent similarities that develop go beyond a matching of the general structure of one part of the body. If two birds take the same ecological role, they will frequently come to parallel each other in behavior as well as anatomy.

Ecological parallels

A bird is a warm-blooded, feathered vertebrate with a beak and four-toed feet. The forelimbs have become flattened elongate wings. The unspecialized bird (*below*) shows the basic body plan on which evolution shapes the adaptations for different lifestyles among the bird groups.

Generalized nonspecialist bird

Scavenging bird

Broad, long wings with wingtip slot "primaries" for efficient gliding

Featherless head and neck

Large beak for tearing skin and flesh

The carrion-eating lifestyle of the vultures and condors demands certain adaptations (*right*) to the basic body plan. The long broad wings allow this type of bird to soar and glide for hours, using minimal energy, as it searches for food.

Powerful feet

Lappet-faced vulture

King vulture

Take, for instance, the example of the niche of a bird that hunts below the surface of the sea on long dives for cuttlefish, squid and fish. Penguins live like this in the nutrient-rich waters around the fringes of the Antarctic continent. Half a world away in the North Atlantic and the Arctic Ocean, some species in a completely different bird family, the auks and murres, hunt in much the same way.

Similar niche roles have provided the template that has shaped the same design in both hemispheres and in both bird families. Each group has a streamlined underwater body shape, with small, tightly fitting feathers. They have sharp, forward-pointing beaks for capturing their prey and pointed paddle-shaped wings for underwater propulsion.

Between dives both types of bird have to swim well at the sea surface, and to accomplish this they both have rear legs positioned far back on the body outline and strong webbed feet. Shared selection pressure has even produced similar body patterning for underwater camouflage.

The scavengers

Large carrion-eating birds also show convergent evolution. Many specialized features of the vulturelike body plan have evolved, apparently independently, in both South America and the Old World tropics. Aspects of design shared by some New World condors and African vultures include a featherless head and neck, easier to keep clean after dipping into a bloody carcass, a powerful beak capable of tearing flesh and skin, and strong talons for dismembering large prey.

Toucans and hornbills are fruit-eating rain forest birds of the New and Old Worlds respectively. Both have massive bills which they use to pluck fruit from the trees.

Although their bills give the birds a superficial resemblance, toucans and hornbills are quite unrelated. They have come to look alike because of their similar lifestyles.

The **Indian white-backed vulture** (*Gyps bengalensis*) thrusts its head into the carcasses it feeds on. Its head and neck are free of feathers which would become clogged with blood and be difficult to keep clean.

Diving, fish-eating bird

Pointed beak for fish capture

Two-tone coloration for underwater camouflage

Flipperlike wings for underwater swimming

Rear-positioned webbed feet for surface swimming

Penguins and auks (murres) share many features, including their two-tone coloration which provides underwater camouflage. The white underside of a penguin makes it hard for potential prey or predators to see from below, while the dark back hides it from those looking down into the dark sea.

Common murre   Galapagos penguin

Plant parallels

Echinops dahuricus (Asteraceae)

Eryngium maritimum (Apiaceae)

Genista hirsuta (Fabaceae)

Spines on leaves and branches protect plants against grazing animals. This useful deterrent has evolved in many different families, including those to which the above examples belong.

The general form of any plant is a result of evolutionary adaptations to particular aspects of the environment. These include climatic factors and the pressures exerted by other living organisms—plants competing for light, water, and space, and animals seeking to exploit the plants as a source of food. Different plants have developed, in parallel, similar forms and strategies for dealing with similar environmental pressures.

The tree is a particularly successful plant form that has evolved separately in a number of different groups. It is an appropriate growth habit where water supply is adequate and winds are not too strong, where temperatures do not fall so low that buds are damaged, and where competition from other plants means that being tall confers a strong advantage.

An adaptable plant form

The tree habit is found in some very primitive flowering plant families, such as magnolias and witch hazels. The very first flowering plants may well have been herbs, and from them developed the whole wealth of today's flowering plant trees. It is a life form also found in the gymnosperms (from conifers to cycads) and is the dominant life form of the group. Tree ferns are still more primitive examples.

Trees were once found even among the lycopods, or club mosses, which are now represented only by short creeping plants. In the days of the great coal-forming swamps of the Carboniferous era, some 300 million years ago, tree lycopods reached a height of 130 ft, similar to that achieved by some rain forest trees today.

Where conventional trees are absent, particularly in isolated locations, such as oceanic islands and the upper reaches of mountains, some rather surprising plant families have evolved members in that form. On the equatorial African mountains giant lobelias, heaths, and groundsels—close relations of the garden weed—have all developed as trees.

Succulents are another type of plant resulting from parallel evolution within different groups. In hot, dry conditions,

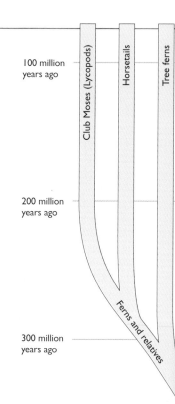

100 million years ago

200 million years ago

300 million years ago

400 million years ago

Club Mosses (Lycopods)

Horsetails

Tree ferns

Ferns and relatives

First land plants

advantages are to be gained by reducing the surface area of a plant to a minimum; this cuts down water loss through the surface pores in transpiration. The best way to achieve this is to adopt a near spherical form. In some plants, such as cacti, the whole stem is reduced to a sphere or a cylinder and leaves may be completely absent or have a short life.

Cacti are naturally restricted to the New World, but there is a living to be made for plants of this form in the dry regions of the Old World. Here the euphorbias have assumed the role and many have cylindrical stems very like the cacti. Groundsels once again display their accustomed adaptability and have evolved species in which the individual leaves are greatly reduced, rounded, and covered in a waxy bloom. A family closely related to cacti, the Aizoaceae, has produced many cactuslike plants in the Old World, especially in South Africa. These include the remarkable living stones (*Lithops*), with

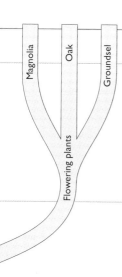

Magnolia
Oak
Groundsel
Flowering plants

The tree form has arisen many times in the course of plant evolution. The fern group began as herbaceous plants, from which the tree lycopods and tree ferns developed. Conifers and cycads are mostly trees, and they may have evolved from a woody fern.

The first flowering plants were probably herbs. Some of these families, such as groundsels (*Senecio*), are mainly composed of small herbaceous plants, but they contain a few species that have evolved the tree habit. These mostly grow on tropical mountains where "conventional" trees are absent.

leaves reduced to the form of pebbles.

One other type that has proved a success and has evolved in a variety of forms is the spiny plant. The predation of animals has led to many species developing protective mechanisms—unpleasant taste, poisons, or stinging hairs—but stiff spines seem to be one of the commonest solutions to the problem of deterring grazers. A whole range of plant families has converged in the evolution of this one form, from thistles to nightshades and even poppies.

In many of the woody species that have adopted the spiny habit, lateral branches have become reduced in growth and come to a sharp point, as in members of the rose family. In other plants, such as thistles, holly, and live oaks, parts of the leaf have become spiny.

Most lobelia species are small herbaceous plants, but tree forms have evolved on high tropical mountains. This example, *Lobelia keniensis* from Mount Kenya in Africa, is about 6.5 ft tall.

Parallels past and present

The physical world is far more constant than the evolving world of nature that fills it. Both in recent times and in more remote geological ages, different organisms have responded in similar ways to similar evolutionary environmental pressures.

All continents normally have a range of environments, some of them similar to those of other continents. Equivalent habitats and niches may be filled by representatives of a single group that has spread throughout the world. The dog and cat families, for example, have dispersed to nearly every continent, and their members have filled the niches for medium to large carnivores.

It also happens that unrelated creatures evolve in parallel to exploit the same opportunities on different continents. For example, ants and termites live in many parts of the world and construct hard earthen nests to house their colonies, eggs, and larvae. They are particularly abundant in tropical regions. Here, then, was an opportunity for the evolution of an ant-eating animal, equipped with strong, clawed limbs to destroy the nest, a long, sticky tongue in a tubular snout to probe for larvae, and a thick, bite-deterring outer covering.

As if to order, four different groups of mammals have independently, but in very similar ways, evolved to fit that niche in tropical zones. They are the anteaters of South America, the aardvark of southern Africa, the pangolin of tropical Africa and India, and the egg-laying spiny anteater of Australia.

Parallels such as this result from similar adaptations in styles of locomotion, feeding, defense, or attack dictated by particular habitats and niches. And since the nature of available niches tends to change little over time, clear similarities can often be demonstrated between living and extinct animals.

Interesting parallels occur, for example, among fast-swimming aquatic animals. These creatures need to be streamlined to minimize water-resistance, they need a powerful tail and paddlelike limbs for propulsion, and a dorsal fin for directional guidance. Their fossils show that the extinct ichthyosaurs and modern sharks and porpoises all evolved exactly these same features.

For rapid running over dry level ground, an animal needs elongated simplified limbs such as those of horses and other present-day ungulates. The extinct ungulate mammals of South America included forms such as *Thoatherium* and *Thesodon* that resembled horses and camels. Similar creatures, such as *Iguanodon* and *Anatosaurus*, existed earlier among the dinosaurs, but they rose onto their hind limbs when running rapidly. Even here there is a living parallel in the kangaroo, which uses its hind limbs alone for rapid locomotion. The kangaroo, though, hops, rather than moves its back legs alternately.

Members of the extinct reptilian pterosaur group once filled the skies. These leathery-membraned flying creatures had wingspans of up to 36 ft. Bats today use similar membranes for flapping flight.

Parallel lifestyles

Feeding adaptations can also produce remarkable parallels. The extinct South American ungulate *Pyrotherium* had a short trunk, much like that of a modern elephant, that enabled it to feed on high vegetation. Giraffes, with their elongated necks, also manage to reach high into trees to feed. An ancient parallel of the giraffe was the long-necked *Brachiosaurus*, one of the largest of the dinosaurs at 75 ft long and 41 ft tall. Like the giraffe, its front legs were longer than its hind legs, and its whole body sloped down from its shoulders.

Even methods of attack and defense produce parallels. The bands of bony armor that protect a living armadillo echo those of the 240-million-year-old amphibian *Peltobatrachus*, as well as the dinosaur *Euoplocephalus*. This dinosaur also paralleled the armadillo's extinct relative *Glyptodon* in having a massive bony club on the end of its tail, which it used to break the bones of its attackers.

Evolutionary pairs

Parallels can be drawn between long-extinct creatures and some modern animals. Although unrelated, they have similar habits and have clearly followed the same evolutionary path and been subjected to much the same environmental pressures.

Animals that can reach foliage high in the trees can exploit a food source unavailable to their shorter competitors. Both the present-day giraffe and *Brachiosaurus* of 150 million years ago evolved long necks that could be used in this way.

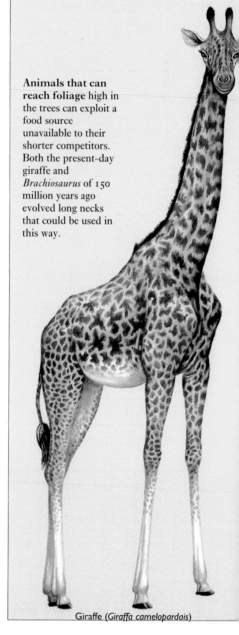

Giraffe (*Giraffa camelopardais*)

The ability to fly opens up many new opportunities. The long-extinct reptilian pterosaurs and the mammalian bats evolved leathery wings made of skin and stretched between the limbs.

Quetzalcoatlus

Spear-nosed bat (*Phyllostomus hastatus*)

The giraffe and *Brachiosaurus* both evolved elongated necks with which they could reach high foliage. *Baluchitherium*, a rhino relative that lived about 25 million years ago in Asia, had elongated limbs instead. Standing more than 16 ft tall at the shoulder, it was the largest land mammal ever known.

Armadillo (*Priodontes maximus*)

Peltobatrachus

Peltobatrachus had protective bony bands and plates on its body very similar to those of today's armadillo. This early amphibian lived in East Africa 250 million years ago.

Brachiosaurus

Changing patterns

The dynamics of living

People often speak of the balance of nature. The implicat
is that the natural world, left to itself, is eternal and tha
is only human beings who bring about changes—and usuall
catastrophic ones at that. This, of course, is not the case. T
Earth and all that is in it has been changing throughout
geological history and is still doing so.

A changing world
The latest major alteration in the world's climate was the
warming that brought the glacial period to an end. Its effect
can still be seen every spring and fall. When, 20,000 years a
the southern margin of the Arctic ice cap lay as far south as
middle of Europe, birds in Africa developed the habit of ma
the short journey north every summer to gather the rich har
of insects that briefly flourished there. As the Earth began t
warm again and the ice to retreat northward, so the journey
birds had to make became longer, perhaps quite suddenly. E
they persisted in the habit and today they may fly 5,000 mil
each spring to reach their summer feeding grounds and repe
the journey back only a few months later. Most of the great
seasonal migrations made by animals today probably origina
in a similar fashion.

It is not only the Earth's climate that changes. So does th
surface of the land itself. Mountain ranges are steadily erode
by ice and water. Lakes fill with the debris of this destructi
and turn first into swamps and then into forest or bogs. The
bites steadily into the coast, undercutting cliffs, and mudban
and deltas built by rivers push out into the sea. Huge areas
forest are regularly devastated by fires. Some of these are in
started by human beings, but many occur naturally, as a resu
of lightning strikes or by volcanic eruptions.

Whole communities of animals and plants specialize in liv
on these changing frontiers moving with them as they move
and exploiting the particular opportunities they bring. When
forest is destroyed by fire, the ground on which it stood is
bathed in sunshine for the first time in centuries. Wind-bor
seeds float down and land on the rich, bare soil that the tree
had created around their roots. There they germinate and
sprout with extreme rapidity. For a few seasons they domina

the clearings, blooming among the charred logs and setting their seed. But then the seedlings of bigger trees, which have grown very much more slowly, rise among them and over-top them. The pioneer plants no longer get the sunshine they need and they die. But their own seeds are now sprouting in other, more recently fire-cleared territory.

In North America one of the commonest of these pioneer plants is called, very appropriately, fireweed. In Europe it is known as willow herb. When, half a century ago, many of the cities of Europe were devastated by bombs, the fireweed found a new kind of seedbed and sprouted among charred timbers that had once been roof beams. So bombed sites everywhere became carpeted in their pink blossoms.

Evolution and survival

Living things not only change their territories in response to a shift in physical circumstance, they are also capable of changing in themselves. They evolve.

Imagine a flock of land-living birds, caught in a sudden storm of hurricane strength and swept out to sea. Before they fell from the sky, totally exhausted, they reached a remote volcanic island. There they landed and their lives were saved. This island inevitably had a much more restricted range of food on it than they had in their original home. They had to swallow unfamiliar food or die. Some failed to do so and paid the price, but others managed to find something edible among the few plants growing on the island, from the spiders and insects that had drifted there earlier, or from the small invertebrates that were washed up on the shore.

The birds had neither the physical strength nor the instinctive route-finding ability to make their way back to the mainland. They were permanently marooned. Those that survived in due course mated, nested, and reared their young. As time passed, certain individuals began to concentrate on particular foods. Some of them had slightly heavier and more powerful beaks and were able to crack seeds that others could not manage. The rewards for doing so were considerable, for since no others could, they had this food resource to themselves, and naturally they concentrated on it. Other birds with perhaps slightly longer beaks managed to reach the nectar that was hidden away in the depths of some of the flowers. Still others, a little more agile in the air, became expert in catching insects. These physical differences in their bodies were inherited by their offspring. As the natural sifting continued over many generations, separate populations eventually developed that were markedly different from the original immigrants. So several new species arose, descended from the original uniform flock.

Interpreting natural history

This scenario may sound glib. But such events are certain to have happened on several occasions. The most famous instance occurred on the Galapagos Islands in the Pacific Ocean, 560 miles off the coast of Ecuador. The wandering ancestral birds were probably South American finches, and arrived there many thousands of years ago. Today, there are more than a dozen different species of them, each with its own diet and physical adaptations that enable it to feed in its own particular way.

Such things also occurred on the Hawaiian Islands. There, too, the ancestral birds are thought to have been finches of some kind, and they gave rise, ultimately, to about 40 different species. Some we know only from fossils, and others have recently become extinct in the face of the major changes that have taken place in the islands' ecology since the arrival of man. But there are still a dozen or so surviving that are known collectively as honeycreepers. And in New Zealand there may have been several such avian invasions, but they occurred so long ago that it is now very difficult to decide what the original ancestral birds were.

These extraordinary happenings on remote and isolated islands may seem to have nothing to do with the main progress of life as it takes place on the great continents. But the very isolation that keeps an island community apart from other influences, and the small number of species comprising it, make the interpretation of its history very much easier than in more complex communities. It was, indeed, a visit to the Galapagos that started the train of thought in the mind of Charles Darwin that ultimately led him to propose a solution to that most crucial question of all in biological science, the origin of species.

Airborne migrations

Sedentary animals are those that are born, grow up, mate, and live out their entire lives within one small area. For an earthworm, this may be only some feet across, for a hedgehog, a few miles. Their niches and patterns of adaptation are tied to the single plot of land where they live, and their capacity for feeding, reproduction, and survival must, therefore, be constrained by the possibilities inherent in their home territory.

Other animals can be highly mobile. This is particularly true of large, fast-running mammal herbivores, of many birds that can fly, and of some winged insects, whose mobility opens up to them a whole new world of niche possibilities and gives them the option of changing their habitats.

The movements may be rather haphazard shifts, brought about by local difficulties or opportunities. For instance, the drying up of a savanna waterhole can induce the birds that feed and drink there to move hundreds of miles to a new water source. In an unpredictable climate, mobility may produce a strategy of wandering movement based on changing local conditions.

Seasonal movements

In contrast, the more patterned type of animal population movement known as migration usually involves a recurrent shuttling between two or more geographical areas in a regular sequence. Almost always, these movements are seasonal, "tuned" in time to the more or less predictable yearly changes in climate in the areas the animals visit.

The most dramatic migrations in terms of total distances traveled must be those of birds and, to a lesser extent, of winged insects. In both the New and Old Worlds, where continents straddle lines of latitude from the cool polar regions to the equatorial tropics, migrating bird species are common.

In most instances, their patterns of migration clearly demonstrate the underlying evolutionary "logic" of the migratory niche strategy: This is an advantageous logic tied to changes in food availability and climate.

A single example, that of the Old World garden warbler, provides many insights into this logic. Like all warblers, the garden warbler feeds on small insects, which it catches among thick vegetation when the plants are in full leaf. It can feed in this way throughout the year by shuttling between Europe and central Africa. So while Europe is locked into winter conditions, which preclude insect feeding for all but the most hardy and opportunistic birds, the garden warblers are feeding in African warmth.

In spring the birds migrate north across the Sahara. Before this journey they feed intensively, laying down supplies of body fat to fuel the energy demands of the long flight to come. They cross the Mediterranean at the eastern end, and spread out through their summer ranges in Europe, where they pair up and breed. In the fall, after again building up fat supplies, the surviving parent birds and their offspring return to Africa, crossing at the western end of the Mediterranean.

By careful matching of flight times with seasonal changes, the birds ensure that they are present in each half of their living space when food supplies are plentiful. Such synchrony is achieved by inbuilt biological clocks in the birds' brains that are influenced by day length. In their summer location, they can sense the passage of seasonal time by the shortening days as fall approaches, while lengthening days in the spring induce hormonal changes that bring the birds into breeding condition at the right time.

The navigational ability of all birds on migration, much of which is instinctive, is a subject of intense scientific enquiry. It probably depends partly on sightings on sun and star positions, partly on patterning of polarized light in the sky, and partly on a magnetic sense that some birds appear to possess. With this sense, they can directly perceive the position of the magnetic north or south pole, and so orientate themselves precisely on the Earth's surface.

The magnificent monarch butterfly (*Danaus plexippus*) has one of the longest of all insect migrations. Five or more generations (eggs, caterpillars, then adults) are needed to complete one migration cycle.

Eastern summer range

Western summer range

Winter roost sites

The insects overwinter in mass roosts in trees in warm southern California or near Mexico City. In spring they migrate north—some even reach Canada by the late summer—then return south for the winter.

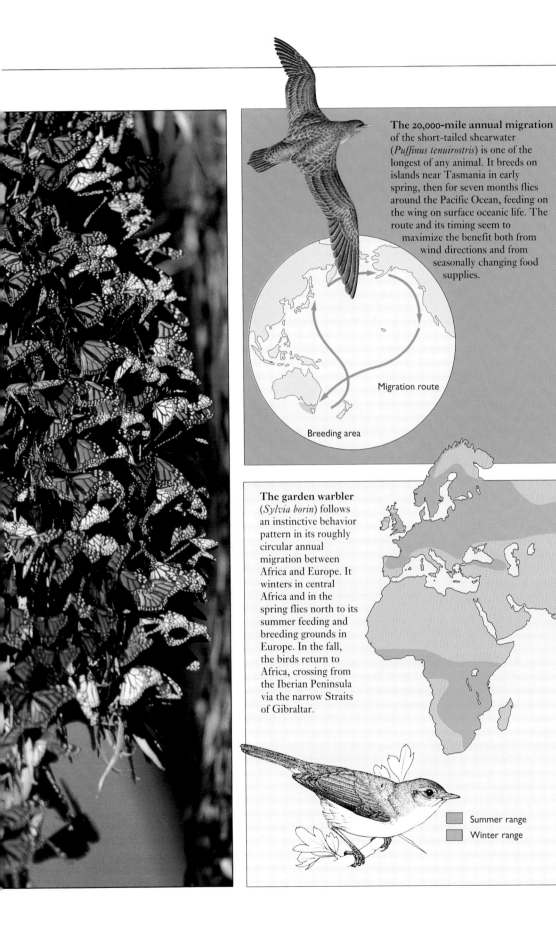

The 20,000-mile annual migration of the short-tailed shearwater (*Puffinus tenuirostris*) is one of the longest of any animal. It breeds on islands near Tasmania in early spring, then for seven months flies around the Pacific Ocean, feeding on the wing on surface oceanic life. The route and its timing seem to maximize the benefit both from wind directions and from seasonally changing food supplies.

Migration route

Breeding area

The garden warbler (*Sylvia borin*) follows an instinctive behavior pattern in its roughly circular annual migration between Africa and Europe. It winters in central Africa and in the spring flies north to its summer feeding and breeding grounds in Europe. In the fall, the birds return to Africa, crossing from the Iberian Peninsula via the narrow Straits of Gibraltar.

Summer range
Winter range

The arctic tern probably holds the record for distance traveled during an annual migration. This slender relative of the gulls flies more than 25,000 miles from its Arctic breeding sites to Antarctic feeding grounds and back again the following spring. Its journey enables the tern to take advantage of the brief but productive summer months of both polar zones.

Overland migrations

The yearly migrations of nonflying animals are usually linked with the availability of food or the animal's breeding cycles, or with both. Such movements are ultimately driven by seasonally changing climate patterns. The word "migration" is normally used only when the distances involved are great, but many terrestrial animals make much shorter annual movements linked with mating and food availability; in essence, these are also migrations.

A common toad, for instance, will make such seasonal movements mainly for reproductive purposes. For most of the year, an adult toad lives in moist vegetation at ground level, feeding on terrestrial invertebrates; but in the spring it will return, over open ground if necessary, to a pond to mate, often the same pond in which it was itself spawned. The distance traveled may be just a few hundred feet, but the movement is, in microcosm, the pattern of any annual migration.

The really impressive land migrations are, however, those of the large herbivorous mammals. Seasonal climatic changes—whether centered on a cycling between hot and cold temperatures or between different levels of rainfall—can make long-distance yearly migrations part of an efficient life cycle for these animals.

In the recent past, huge herds of North American bison, numbered in tens of millions, cropped the Plains grasslands from as far north as Alberta to New Mexico in the south. Each spring the bison moved north in the lengthening days and milder weather, giving birth to their young as the lush, relatively protein-packed spring grasses sprouted. As cold northerly winds began to blow in the fall, the bison herds moved south again to escape the harshest winter conditions. These migrations ceased after the depredations of nineteenth-century European hunters took the bison almost to the point of extinction.

Today the caribou show a similar life plan to that of the bison. They spend the winter in the deep coniferous forests of Alberta, Saskatchewan, and Manitoba in

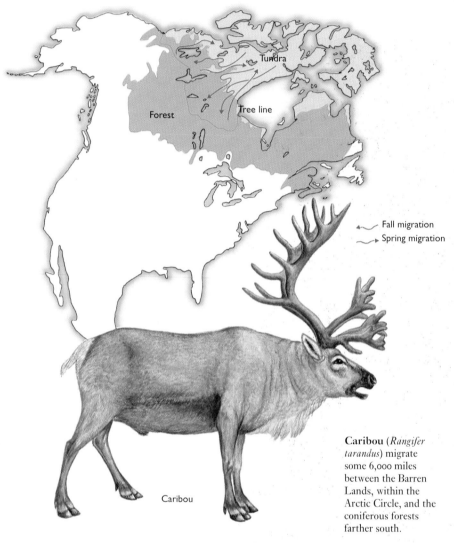

Fall migration
Spring migration

Caribou

Caribou (*Rangifer tarandus*) migrate some 6,000 miles between the Barren Lands, within the Arctic Circle, and the coniferous forests farther south.

Canada, and in the spring begin a northward trek to the Barren Lands close to its northern coast.

On the journey north, the young caribou are born, and within hours of their birth are running with the rest of the herd. The nutritional "payoff" for the vast expenditure of energy on the migration is fresh, good grazing. In the brief Arctic summer, there is suddenly a huge crop of rapidly growing grasses, and as the snow cover melts, edible lichens are exposed. The caribou fatten up and mate in the late summer, before the return—southward—migration begins.

In the tropics, large seasonal temperature changes are unknown; instead, seasonally changing rainfall is often the norm. And it is the changes in vegetation caused by the varying rainfall that trigger the yearly migrations of East African plains herbivores such as the wildebeest. These movements have been carefully monitored in the Serengeti National Park in Tanzania, where wildebeest herds make complex migrations linked to the availability of croppable grasses.

As the midequatorial rain belt oscillates across the Serengeti with a yearly cycle, it produces a marked wet season from January to March and a dry season from June to September. The migrations of the wildebeest, often in mixed herds with zebra, topi, and gazelles, take them on an approximately circular path that is dictated by the availability of grazing.

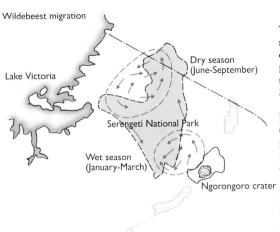

Wildebeest migration

Lake Victoria

Dry season
(June-September)

Serengeti National Park

Wet season
(January-March)

Ngorongoro crater

**The annual
movements of the
equatorial rain belt**
produce a wet and dry
season in the
Serengeti.
   In the wet season
herds of wildebeest
(*Connochaetes taurinus*)
graze in the southeast,
near the Ngorongoro
crater. In April/May
the herds move north
and west to feeding
grounds nearer Lake
Victoria, where they
spend the dry season.
Around October, the
return journey south
begins.

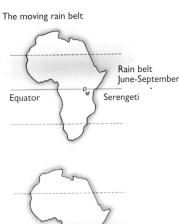

The moving rain belt

Rain belt
June-September

Equator   Serengeti

Equator   Serengeti

Rain belt
January-March

Marine migrations

Any migrating land animal that cannot fly eventually comes up against the insuperable barrier of the sea. No comparable difficulty exists for migrating marine animals, since the oceans of the world are a continuous, connected whole. This immense potential for movement has had two major consequences. First, large numbers of highly mobile sea creatures are distributed worldwide; second, long migrations are extremely common.

Migrations in the sea are linked with the same imperatives as those on land. Salmon, eels, turtles, and whales migrate in order to maximize their chances of survival (through feeding) and of reproductive success (through breeding).

Not all marine migration patterns are locked into an annual cycle; the migrations of some species conform to much longer time scales. Those of the European and American eels are a well-researched example. The 16 eel species (genus *Anguilla*) of the world are all catadromous; that is, they breed in the sea but feed and grow in freshwater.

The migrational habits of eels were unknown until the early part of this century. From earliest times, the fact that in Europe adult fish are common in lakes, streams, and rivers, but no eggs or young are ever found, had caused great speculation. The ancient Greek philosopher Aristotle even suggested that they were spontaneously generated in the mud at the bottom of lakes. It took the determined work of the Danish oceanographer Johannes Schmidt to unravel at least the main components of the mystery.

His findings from the 1920s onward have shown that both American and European eels spawn in the Sargasso Sea, hundreds of miles due east of Florida. Here, at depths of 1,300–2,500 ft, adult eels produce their eggs and fertilize them. The tiny, transparent larvae that hatch from these eggs are carried by the ocean currents, the American form up toward the east coast of the United States, the European on the much longer Gulf Stream route to Europe.

During this drifting migration, which can take three or four years, the larvae slowly metamorphose into leaf-shaped, flattened leptocephalus larvae that feed on plankton. When they reach the shallow waters of the continental shelf, they change into glass eels, then elvers. In this form they move into river estuaries along the entire western coastline of Europe.

In freshwater habitats the eels feed and grow—for more than 10 years. When full adult size has been reached, they change color, from greenish to silver, grow larger eyes, become sexually more mature, and move downriver back to the sea. It is assumed that they then make the return migration to the Sargasso Sea to breed, but no eels have ever been caught in mid-Atlantic on this journey. This has led some scientists to suggest that they do not return from Europe and that all the European eels are derived from fertilized eggs of the American stock. Certainly the two species are extraordinarily alike.

The green turtle's journey

A cosmopolitan marine reptile of tropical waters, the green turtle is found only where the sea temperature is above 68°F. Like most turtles, it spends the greater part of its life at sea but still has to haul itself ashore on a sandy beach to lay its eggs. These breeding beaches are always highly traditional locations adopted by particular populations of this turtle.

One amazingly remote green turtle breeding site is on Ascension Island, in the mid-Atlantic. This tiny speck of rock is the egg-laying site for a population that lives in the warm, shallow waters off the east coast of Brazil, more than 1,240 miles to the west. For most of the year, the turtles feed here on algae and eelgrass; then, by methods that are not fully understood, they navigate their way to Ascension. After a period of offshore courtship and mating, they lay their eggs, in December, on the beaches of volcanic sand. After 10 weeks' incubation, the baby turtles hatch synchronously at night and scramble down to the sea. Nothing, however, is yet known about the early life and movements of these hatchlings.

□ Eel Breeding ground   □ Eel Migration path

European eels (*Anguilla anguilla*) spawn in the Sargasso Sea east of Florida. The larvae hatched from these eggs slowly drift across to Europe, a migration taking several years. Eventually they enter estuaries and move upriver where they feed and mature. American eels (*A. rostrata*) breed there too but have a shorter journey to the rivers of eastern North America.

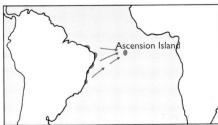

Feeding grounds

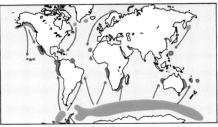

Summer feeding grounds ◼ Winter breeding grounds

The first observations of the migration of baleen whales, such as the humpback, were made more than a hundred years ago. Whalers hunting these mighty sea mammals discovered that they were to be found in quite different latitudes at different times of the year.

Green turtles (*Chelonia mydas*) occur in warm seas worldwide. Those that live and feed off the coast of Brazil make an extraordinary migratory journey to Ascension Island in the Atlantic Ocean to mate and lay eggs. The adult turtles then return to their feeding grounds, leaving their young to incubate, hatch, and find their own way to the sea.

Humpback whales (*Megaptera novaeangliae*) occur in both northern and southern hemispheres and all follow an annual migration pattern. In summer the whales feed in the krill-rich waters of polar regions; in winter, they move closer to the equator. Here, in warmer waters, they give birth to young conceived the previous year, and then mate again.

Traveling the world: animals

Migrations are repetitive, usually seasonal, changes in the geographical position of a population of animals. There are, however, other, fundamentally different types of population movement that are neither seasonal nor repetitive. These are range expansions in which a species extends its geographical range steadily, generation by generation, in one or more particular directions.

Over thousands or millions of years, all animal species will expand or contract their ranges as a result of their relative success or failure. The alterations will be due either to changing patterns of competition or to changing environmental conditions. For all their assumed ubiquity, these long-term alterations can often only be inferred from present-day distributions. Direct evidence of range expansion in today's species is rare and is most evident in highly mobile animals, particularly bird species.

The fact that direct evidence is limited to the last 100 years or so raises a problem of interpretation. Any alteration over this period will have occurred at a time during which the human population has been rising exponentially, and when there has been immense human interference with habitats and conditions over more and more of the Earth's surface. This makes it difficult to decide whether a range alteration has been "natural" or largely or partly caused by humans.

Some expansions appear to have taken place without human interference being a major contributory factor, as the examples of two bird species, the collared dove and the cattle egret, clearly demonstrate.

Prior to 1930, the collared dove—a small, grain-eating member of the pigeon family—was essentially an Asiatic bird, extending no farther into Europe than Turkey and adjacent areas. Then it commenced a remarkably speedy extension of its range in a consistently northwestern direction. It has now colonized the whole of central Europe and much of Scandinavia and Britain. It has recently reached Florida as a result of human transport and may well be on the point of an expansion of range in North America.

Ornithologists and ecologists do not really understand what has caused the steady northwestern expansion of the species. It may be related to the slowly warming climate of northern Europe or to a lack of specific competitors for the particular niche of the collared dove in these more northerly zones. Or it could be associated with a sudden genetic alteration in a population of doves in the region of Turkey in the 1920s.

The cattle egret

An even greater expansion has occurred in the range of the cattle egret since about 1930. Until then, these beautiful white relatives of the herons were found in tropical and subtropical areas, from Africa to southwest Europe and through Asia to

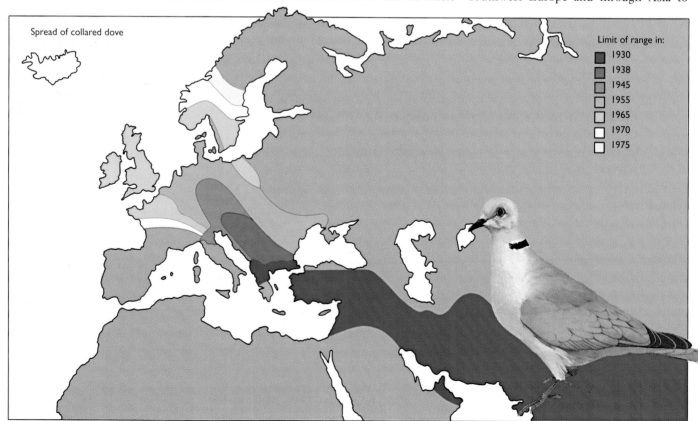

Spread of collared dove

Limit of range in:
- 1930
- 1938
- 1945
- 1955
- 1965
- 1970
- 1975

Japan. Over this already extensive range, they fed on insects, frogs, lizards, and small mammals disturbed by the passage of herds of wild buffalo, wildebeest, and eland, as well as of domestic cattle.

In the 1930s the egrets established themselves as a successful breeding species in South America, having apparently made the Atlantic crossing unaided. They then moved into the islands of the Caribbean, and from there to North America, where they first bred in 1953.

On the other side of the world, at almost the same time, they moved into north Australia from Asia. They have now spread into the rest of Australasia and even to the Hawaiian island chain. These last expansionary movements have, however, been helped by deliberate human introductions. Indeed, the whole of the cattle egret's success is certainly underpinned by the expansion of the range of farmed cattle herds around the world—human influence is never very far removed.

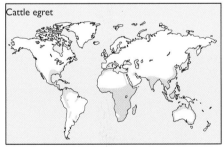

Cattle egret

☐ Distribution before 1930
☐ Present distribution

The successful range expansion of the cattle egrets (*Bubulcus ibis*) is partly due to their adaptability. As well as following wild herbivores to feed on the insects and small creatures disturbed by their movements, they also accompany domestic cattle.

The collared dove (*Streptopelia decaocto*) (*left*) was originally a native of eastern and southern Asia but has made a spectacular range expansion since the 1930s. It has now colonized much of Europe, including Scandinavia, and the British Isles.

Traveling the world: plants

Moving from one place to another is harder for plants than for animals, simply because most plants are literally rooted to the spot. Some plants can spread themselves to another location vegetatively, by fragments that break off and are transported by wind, water, or animals. But it is only at the seed or spore stage of their life history that most plants have the opportunity to extend their boundaries.

Plants are remarkably good at getting from one place to another in seed form and may travel by wind or water. Seeds of cotton sedges, for example, with feathery tufts that keep the fruit airborne, have reached the newly formed island of Surtsey, near Iceland, on the wind. Coconuts have thick, fibrous, air-retaining husks around their fruit, so they float well and can travel by sea. Tropical bean seeds also float well and arrive regularly on the west coast of Ireland from the Caribbean.

Plants may make use of animals' mobility in distributing themselves. Seeds may be carried on fur, or inside the digestive system. Seeds that travel in an animal's gut must survive immersion in digestive juices but may be transported a long way if the creature is migrating.

Humans provide some of the most reliable means of seed transport. Many weeds have adopted humankind as an agent for travel and a provider of sites for survival at new destinations. They may simply hitch a ride. The North American oval sea rocket (*Cakile edulenta*), native to the Atlantic coast and Great Lakes, was accidentally carried to Melbourne, Australia in sand ballast in 1863. By 1882 it had moved on to San Francisco, and today has colonized the Pacific coast of North America, having traveled right around the world.

The climate factor

When their journey is completed, seeds must germinate and establish themselves if their travels are to prove successful. In this they are restricted by factors such as climate. The tropical seeds that arrive on the Irish coast have traveled in vain: it is too cold for them to survive.

When the climate is changing and getting warmer, plants are presented with new opportunities to spread. Hazel, for example, moved rapidly from Spain and southern Europe through the rest of Europe at the end of the last glacial 10,000 years ago. Its nuts were carried by sea, by small mammals and birds, and possibly also by Mesolithic hunters.

When a climate gets colder, warmth-dependent plants, such as hazel, begin to contract in their range. Retreat may result from the sudden death of plant populations in severe weather, or come about gradually through plants' inability to set good seed or through the death of seedlings. The population then fails to regenerate and begins to decline.

Extension or retraction of a plant's range need not be related to climatic change. In Australia a native tree, *Pittosporum undulatum*, recently began extending its range, moving westward in Victoria. This may have been induced by alteration in land use, caused by the suppression of bush fires, or it may relate to the European blackbird—an introduced species that is efficient at spreading the tree's seeds.

It is not always possible to know whether the spread of a plant species is due to climatic change or to the provision of new habitats through human management of the land. In Europe, both spruce and beech have spread westward in the past 3,000 years. The climate has changed during that time, but trees may be taking advantage of changed soils and reduced competition caused by human activity.

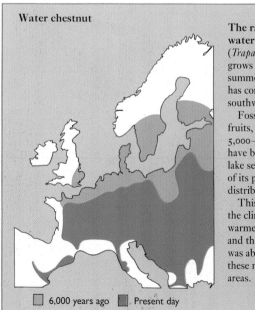

Water chestnut

The range of the water chestnut (*Trapa natans*), which grows best where summers are warm, has contracted southward.

Fossils of its large fruits, dating back 5,000–7,000 years, have been found in lake sediments north of its present distribution limit.

This suggests that the climate was warmer at that time, and the water chestnut was able to grow in these more northerly areas.

☐ 6,000 years ago ◼ Present day

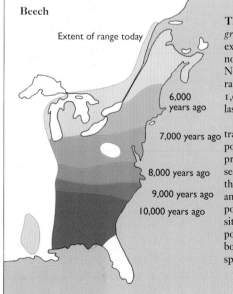

Beech

Extent of range today

6,000 years ago
7,000 years ago
8,000 years ago
9,000 years ago
10,000 years ago

The beech (*Fagus grandiflora*) has extended its range northward in eastern North America at a rate of more than 1,000 ft a year over the last 10,000 years.

Its spread has been traced by analyzing its pollen grains, preserved in lake sediments. By dating the first occurrence in any quantity of beech pollen at various lake sites, it has been possible to ascertain both the course and speed of its expansion.

Seeds of the spear thistle (*Cirsium vulgare*) drift away from the seed head to be dispersed by the wind.

Population explosions

Some animal species alter their location by yearly migrations or steady range expansions, and these movements are reasonably predictable. Quite different are the sudden, unexpected increases in animal numbers that induce unpredictable moves to new areas. This type of change is known as a population explosion, or epidemic, and the movement it causes is an irruption, or invasion. To understand what underlies such irregular events—to find out why locusts swarm or lemmings go on the march—it is necessary first to look at less eventful lifestyles.

An annual census of the numbers of an animal species in a particular area at a specific time of year will usually reveal an overwhelming impression of population stability. Although there may be random fluctuations from year to year, there is usually no consistent trend toward either an increase or decrease in population size. This is despite the fact that all animals have the potential for increasing their population size dramatically by breeding.

Controlling factors

It has been suggested that population size is "regulated," or kept under control, by the links between the interacting organisms in a habitat. Of special importance are those in which one organism (animal or plant) is the food of another, or in which one organism (germ or parasite) acts to damage or kill another (the host). So it would seem that exploding populations are those in which a major change occurs in these or other controls.

Lemmings—small rodents of northern Europe, Asia, and North America—provide an almost mythological example of population explosion. Stories of mass suicide migrations are almost always without foundation. It is true, however, that in most parts of the lemming's distribution there is a locally synchronous rise and fall in population size with an approximately four-year cycle. And at the peaks in this cycle, lemmings disperse widely in unusual numbers, as competition for feeding and breeding space becomes acute.

The main reason for these changes

appears to be their huge reproductive potential. Because lemmings are active and feed in burrows under the snow in winter, a female is capable of producing up to eight litters of three to nine young each year. This potential population explosion is normally kept under control by food and climate restrictions and, to a lesser extent, by predators. But a rapid rise in numbers can result from a long summer, followed by a winter that is neither so mild that the snow melts, destroying the feeding tunnels in the vegetation under the snow, nor so cold that the young die. The resulting competition for food and space then makes for a rapid reduction in numbers.

An example of a mammalian population cycle where predation may play a more central role is that of the lynx and the snowshoe hare in northern Canada. Both species, one a hunting cat, the other its main prey item, have approximately 10-year cycles of numbers. It seems that the hares' expansions result from changes in climate and the availability of food. The contractions are partly due to increased competition but also to increased predation by the expansion of the lynx population, itself caused by the extra hares.

An insect plague

Desert locusts show population explosions

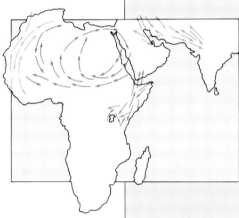

When desert locust populations expand but food is limited, the developing insects change into a voracious migrating form of the species. Huge swarms (*left*) move along generally predictable routes in Africa and the Middle East.

Trapping records, going back to the 1850s, show that fluctuations in the population density of both the lynx and the snowshoe hare follow an approximately 10-year-cycle. Numbers in peak years are more than 10 times higher than in the intervening troughs.

The immense reproductive capacity of the Norway lemming is partly the result of its extremely short gestation and maturation periods. Gestation lasts only about 20 days, and the three to nine young are weaned within three weeks. Two weeks later they, too, are capable of breeding.

In ideal conditions, the total generation time (from mother mating to daughter mating) is only eight weeks.

at unpredictable intervals. These have huge economic consequences in parts of Africa and Asia where they occur, since migrating swarms of locusts can destroy the crops over great tracts of country.

When food is plentiful, the desert locusts behave like normal, nonmigratory grasshoppers. If, however, increasing numbers of changing climatic conditions reduce their local food supplies, the developing locusts may transform into a swarming migratory type, with long wings and a brighter coloration. This transformed locust is active by day rather than by night and are the notorious Old Testament ravagers of crops.

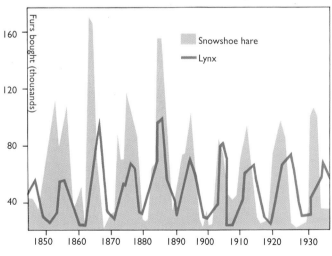

Colonizing islands: getting there

Oceans are the most effective barrier to the dispersal of land organisms and the colonization of distant islands. Any animal that can swim 1,000 miles will be so thoroughly aquatic that it could not adapt to a terrestrial existence; relatively few plants, such as the coconut palm and sea beans, have fruit or seeds that can survive a long period of immersion in saltwater.

Occasionally land animals and large plants are able to reach islands on masses of floating vegetation washed down by tropical rivers after heavy storms, and small animals or their eggs could easily be carried in this way, too.

But it is much more likely for the seeds of plants and even intact small animals to make an aerial ocean crossing. The tiny (.0004–.04 in) spores of ferns and mosses are easily carried by the wind, and often form an unusually high proportion of the flora of remote islands such as Hawaii.

Seed dispersal

Many flowering plants have adaptations that ensure their seeds are carried away from the parent. Members of the Asteraceae family, such as the dandelion and thistle, have feathery tufts on the seeds, allowing them to drift freely with the wind, and this family is highly successful in crossing oceans.

Bidens, another member of the Asteraceae, is extremely widespread in the Pacific islands because its hooked seeds can fasten to bird feathers. Other plants hitch a ride by having seeds with sticky secretions that attach them to the fur or feathers of mammals or birds. Tiny snails and their eggs can also be carried by birds, in the dried mud on their legs and feet.

Island floras also contain an unusually high proportion of plants with fleshy fruits. Once eaten by birds, their seeds may be carried long distances before they are excreted and can germinate. Blueberry, sandalwood, mint, lily, and nightshade probably all arrived in the Hawaiian islands in this manner.

The dispersal of plants is basic to the dispersal of all other living organisms to the islands; the more plants there are, the more ecological niches there will be for animals to occupy. For example, the number of bird genera on islands closely reflects the diversity of the plants among

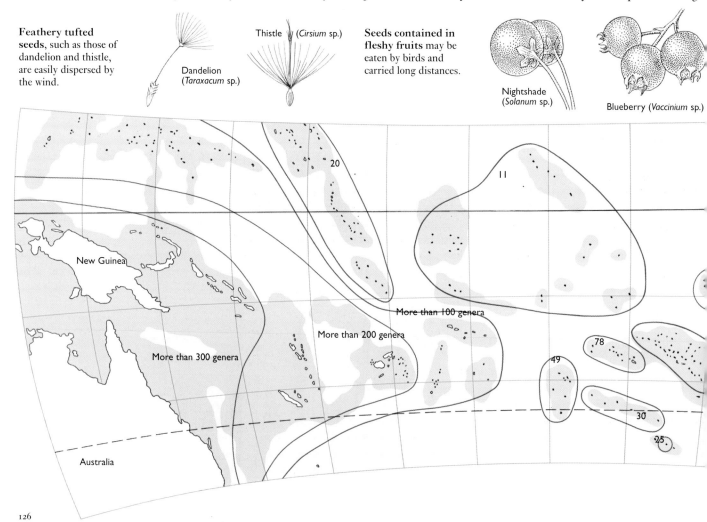

Feathery tufted seeds, such as those of dandelion and thistle, are easily dispersed by the wind.

Dandelion (*Taraxacum* sp.)

Thistle (*Cirsium* sp.)

Seeds contained in fleshy fruits may be eaten by birds and carried long distances.

Nightshade (*Solanum* sp.)

Blueberry (*Vaccinium* sp.)

New Guinea

Australia

20

11

More than 100 genera

More than 200 genera

More than 300 genera

.78

49

30

25

which they live and upon which many of them feed.

Much can be learned by studying what happens when an island has been totally devastated and is then gradually recolonized. The most dramatic natural example resulted from the explosion in 1883 of the island volcano of Krakatoa in the East Indies. Twenty-five years later, 13 species of bird had colonized the island. After another 10 years there were 31 species of bird, but two of those previously found had disappeared. Another 10 years on there was the same total number of species, but another five species which had earlier colonized the island had vanished.

So it seems that there must be a balance in nature between the rate at which new organisms colonize an island and the rate at which previous colonists become extinct. At first, many species can colonize a newly appeared island, but as time passes the rate of appearance of new species will gradually drop. At the same time, the rate of extinction of existing species will rise, partly because the more there are, the more there are at risk, and partly because the competition between them becomes greater.

Competition between species also leads to more specialization, and this very exclusivity means smaller populations and greater vulnerability. But eventually a balance will be arrived at, and unless conditions alter dramatically, the number of species will now change little, even though new ones may appear and others become extinct.

Coconut palms grow on tropical beaches just above the high water mark, and their nuts are easily washed out to sea. They float well and can survive months of immersion before reaching a new beach, where they land and, if temperatures are right, germinate. Hence the coconut palm has been highly successful at dispersing itself around the tropical world.

Beggar's-tick
(*Bidens* sp.)

The hooked seeds of plants such as beggar's-tick cling to bird feathers and are widely dispersed.

Contour lines for the number of flowering plant genera found on Pacific island groups are shown on the map. East of the 100 genera line, figures for each group are given. The farther east the island, the fewer the plants that have succeeded in reaching it.

Equator

Tropic of Capricorn

24

Colonizing islands: staying and surviving

The diversity of an island's flora and fauna is affected not only by the rate at which new colonists arrive and old ones become extinct but also by its area and topography. A great deal can be learned of the nature of island biogeography by studying the history of a single island such as Barro Colorado in Central America.

Barro Colorado is about 6 sq miles in area and is separated from the nearest mainland by stretches of water at least 1,000 ft wide. Though it lies in the tropics and has an average rainfall of 100 in a year, it also has a dry spell from January to March when less than 4 in falls.

This small island is covered in semi-deciduous tropical forest, in which some trees lose their leaves in the dry season, others only in unusually dry years. The forest is diverse and fairly dense, with each 2.5 acres containing 50–65 species of tree, and about 170 trees with a trunk diameter of more than 8 in, as well as many more smaller ones. In the south-western part of the island, trees grow to heights of 100–130 ft and are 200–400 years old; in the northeast, the trees are only 65–100 ft tall and are probably only about 100 years old.

Animal extinctions

A number of mammals and birds have become extinct since the island formed, and there seem to be three reasons for this: the size of the island, changes in vegetation, and hunting by humans.

Barro Colorado is too small to support a big population of large animals, and a small population is always at risk of extinction; the mountain lion and white-lipped peccary probably disappeared for this reason. Similarly, animals with very specialized feeding patterns are unable to find their preferred food sufficiently often on so small an island. As a result, the ocellated antbird, which feeds only on insects flushed out of the forest litter by swarms of army ants, has been unable to survive, although the spotted antbird, with less restricted feeding habits, flourishes.

When the island formed, there were still some agricultural clearings in which lived rodents such as the pygmy squirrel and pygmy rice rat. There were also old clearings that the forest was still recolonizing, in which there lived 32 species of birds, such as wrens, woodpeckers, fly-catchers, doves, and species of hawk and falcon that preyed on them. All these animals and birds are now extinct because their environments have disappeared.

Dangers of island life

Hunting by humans before the island became a reserve led to the extinction of the spider monkey and the tapir. Both have now been reintroduced, but their numbers are still so low that they may again become extinct, as may the ocelot. The extinction of the mountain lion and the small numbers of hunting ocelot meant that their natural prey, such as monkeys, coatimundis, and opossums, became more common. The competition they provided led to the extinction of several species of birds that nest or forage on the ground.

In addition to identifying these long-term changes in the flora and fauna of the island, scientists have also been able to monitor the effects of an unusually rainy dry season. Many plants require the stimulus of a dry season before they can flower, and in 1970 the dry season was so wet that it failed to trigger that flowering in many cases. Less than half the normal number of species bore fruit, and that of others was much reduced, so the total crop was only one-third of its usual level. The birds and mammals relying on this fruit for food suffered badly, and despite turning to unaccustomed food sources, many did not survive. Coatimundis, agoutis, collared peccaries, opossums, howler monkeys, armadillos, and even the occasional tree sloth were found lying dead in the forest. None could emulate the fruit-eating birds, such as parrots, parakeets, and toucans, which simply flew away.

All this confirms and underlines the basic danger of life on islands, where a small area provides less variety of environment. There is less chance of avoiding the occasional but inevitable ill-fortune, and little chance of escape.

Barro Colorado

This small island was cut off from the mainland during the construction of the Panama Canal, when river waters were dammed and rose to form Lake Gatun. The lake itself is some 80 ft above sea level.

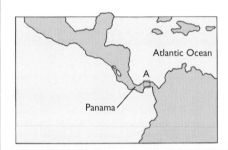

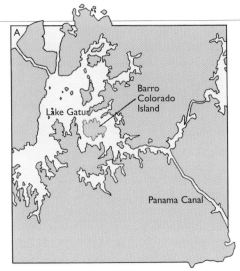

Pollen studies show that forest covered Barro Colorado 35,000 years ago. But from 7000 BC much of this was cleared and corn was grown until 1500–1600, when the local Indian population was decimated by the Spanish conquest. What is now the older forest may then have started to grow in the abandoned clearings. Today's younger forest may date from a reduction in local farming about 1800, when the heavy traffic through the area of gold seekers on their way to California ceased.

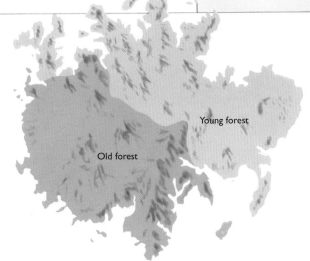

Barro Colorado lies in the Panama Canal area in Central America. The island has been studied since 1916 and in 1923 became a nature reserve and natural laboratory linked to the Smithsonian Institution in Washington, DC.

Isolated populations

As land-living animals, we naturally think of isolation as resulting from water surrounding a piece of land. But aquatic animals become isolated if the body of water in which they live is surrounded to form an inland sea or fresh-water lake. In this respect, the Great Lakes of East Africa are of particular interest.

As in so many other instances, plate tectonics has set the stage upon which biological events have taken place, for East Africa is an area containing many rifts, which continue up toward the Red Sea, where an oceanic ridge is beginning to separate Arabia from Africa. Some of these rifts, formed 1.5–2 million years ago, created the deep, elongate lakes of Tanganyika (4,820 ft deep) and Malawi (2,300 ft deep). Lake Victoria is both younger (750,000 years) and shallower (330 ft), but all three are notable for containing an enormous variety of fishes belonging to the cichlid family.

A wealth of species

Cichlid fishes are found in streams, rivers, and lakes in both Africa and South America, but nowhere do they appear in as much diversity as in the Great Lakes of Africa. Lake Malawi contains 200 species, Victoria 170 species, and Tanganyika 126; and nearly every one of these species is found only in one particular lake and must have evolved there.

The evolution of a new species is believed to take place when a population is isolated from its relatives. Only then can it develop the features that allow it to follow a new way of life and evolve the genetic differences that prevent it from mating with its former relatives and merging back with them. So it is not surprising to find that the fishes of each lake are different. What is at first puzzling is how so many species could have evolved in each.

The history of Lake Victoria appears to be different from that of the other two. It is far shallower and seems to have formed when the land to its west rose, damming the flow of several rivers that had previously drained westward into the Zaire

river system. Each of these rivers would have contained its own species of cichlid fishes, and as each river swelled and expanded to form a small but deepening lake, new environments would have been created, so new species of fishes could have evolved to make use of them.

Evolution might have stopped there had the new Lake Victoria remained constant; but changes both in the shape of the land in this geographically unstable area and in the climate would have caused the size and shape of the lake to alter. Sometimes it would have shrunk to a series of separate lakes and lagoons, in

each of which new species could evolve; then later earth movements or heavier rainfalls would have reunited the lakes and their new types of fishes.

This is borne out by the fact that five new species of cichlid fishes have evolved in a small lake of 12 sq miles, which has been separated from Lake Victoria for only the last 4,000 years. When land subsidence or flooding reunites the two lakes, these five new species will become part of the fauna of Lake Victoria.

In the narrow, deep lakes of Tanganyika and Malawi, it seems more likely that new species evolved from populations that

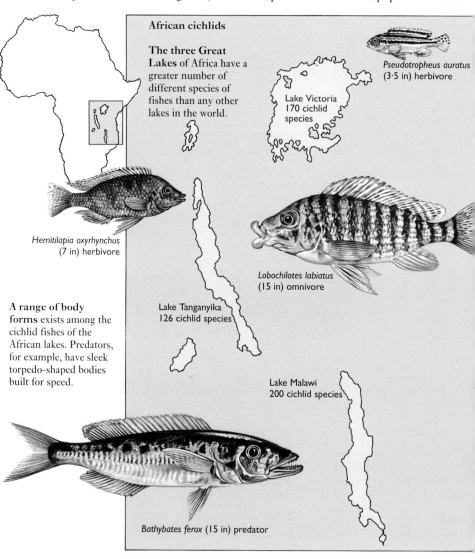

African cichlids

The three Great Lakes of Africa have a greater number of different species of fishes than any other lakes in the world.

Pseudotropheus auratus (3·5 in) herbivore

Lake Victoria 170 cichlid species

Hemitilapia oxyrhynchus (7 in) herbivore

Lobochilotes labiatus (15 in) omnivore

A range of body forms exists among the cichlid fishes of the African lakes. Predators, for example, have sleek torpedo-shaped bodies built for speed.

Lake Tanganyika 126 cichlid species

Lake Malawi 200 cichlid species

Bathybates ferox (15 in) predator

became isolated because of their habits. Every area of rocky or sandy bottom has its own population of cichlids, each of which is a potential new species.

This type of evolutionary change may also occur in cichlids because of their social and mating behavior. The males of different species are often distinguished by their bright colors and patterns: gold and black horizontal stripes, royal blue and black vertical bars, patches of bright blue or orange. These colors are shown off in territorial mating displays.

Certainly, none of this evolutionary radiation would have been possible had the cichlid fishes not been able to adapt to a variety of foods and environments. In addition to ordinary teeth along the edges of their jaws, they have evolved a battery of teeth on the bony skeletal supports to their gills. The shapes and numbers of the teeth are adapted to the type of food each fish eats, as is the shape of the jaws.

Such adaptations, together with a considerable range in length (1.5–31.5 in) and the colonization of both rocky and sandy shallow shores, open waters, and depths up to 100 ft, have allowed as many as 200 species of cichlid to coexist in these lakes in adjacent but separate "species flocks."

Lake Tanganyika is the seventh largest lake in the world, with an area of some 13,000 sq miles. Like the other two great African lakes, this "watery island" has its own unique fauna.

Petrotilapia tridentiger
Feeds on algae scraped from rocks

Lethrinops gossei
Feeds on invertebrates dug from sea bed

Piecodus paradoxus
Feeds on scales of other fishes

The human
impact

Forces of
the future

Human beings inflict huge changes on their environment. So, of course, do rabbits. Indeed, since all animals by th[ei]r very nature depend directly or indirectly on plants, and almo[st] all interact in some way with other animals, the disappearance or the increase of *any* animal species affects the pattern of life within a biological community.

But human beings have two characteristics that distinguish them from all other animals. First, they have, over the past fe[w] centuries, increased in number so swiftly that today they are [by] far the most numerous and the most widely distributed of any large animal species. Second, they have become so technologically inventive that they can fell an entire forest in [a] day, create a new species of living organism to suit their own requirements, and pulverize a city and all life within it in an instant.

For centuries this most powerful of species has exploited th[e] natural world for its own ends. Human beings operate as thou[gh] the planet's resources are inexhaustible, believing that if they want any more of anything they can take it.

The dodo has the sad distinction of being one of the first o[f] the world's animals known to have been annihilated by this process in modern times. It was a giant flightless pigeon that lived on the island of Mauritius in the Indian Ocean. Europea[ns] did not discover the island until the sixteenth century, but thereafter sailors made a point of landing on it in order to clu[b] the defenseless birds on the head and have a rare feast of fres[h] meat. Within less than 200 years, every single dodo had been killed.

In succeeding centuries, more animals followed the dodo. European settlers in South Africa found vast grassy plains thronged with huge herds of antelope and wild relatives of horses. They hunted them for food and for fun. Within a few decades, they had destroyed the majority of the herds. One species, a kind of half-striped zebra called a quagga, was total[ly] exterminated. Others survived only in greatly reduced number[s] farther north in wilder country.

The pattern of destruction continued as more of the Earth's surface was claimed by human beings for settlement or exploitation. By the 1950s naturalists discovered that at least 1 mammals and nearly 200 birds, let alone other kinds of animals and plants, were on the verge of extinction. Appalled at this prospect, they banded together, raised the alarm, and started collecting money to try to protect the survivors. Some species, reduced to only a handful of individuals, had to be taken into captivity in order to encourage them to breed and give them protection while they were doing so. Tracts of wilderness that were the last refuge of particularly endangered animals were turned into reserves and given legal protection against further despoliation.

The rain forests

But today it is not just individual species of animals and plants that are in danger of disappearing, nor even precious remnants of what were once great wildernesses. In the last few decades, human beings have created changes on a global scale. We are interfering with the circulation of water between the Earth and the atmosphere by demolishing a key link in that process, the rain forests. For millions of years, these have acted as reservoirs, absorbing the deluges of the rainy season and releasing the water steadily into the rivers during the dry. Now we are cutting down those forests, leaving in their place either bare, ravaged earth or fields planted with crops that, in nearly all cases, fail after a few seasons. In both instances, the land becomes covered with a thin scrub that cannot retain the water, either in the substance of its wood and leaves, or in the soil, which is only feebly held together by the thin roots. So the soil erodes and the land becomes alternately ravaged by floods and baked dry into a desert. The moisture that once rose in such quantities from the forest canopy to form clouds no longer does so. The whole pattern of rainfall over the continent is disrupted.

We are changing the global climate in another way, too. Our industries, for the past two or three centuries, have drawn their power from carbon compounds that accumulated deep in the Earth's crust as oil and coal. When we burn these fuels, we produce from them a gas, carbon dioxide. This accumulates in the atmosphere and forms a screen that acts like a pane of glass in a greenhouse. It allows the sun's rays to pass through it, but it reflects back the heat those rays produce when they strike the Earth. As a consequence, the whole planet is now warming. If this process continues unchecked, then the polar ice caps may melt, causing sea levels to rise, and deserts may expand. Felling forests will further add to the carbon dioxide in the atmosphere.

These changes are likely to occur swiftly. A few species of animals and plants may be able to tolerate them. Some may succeed in changing their distribution patterns and so keep pace with the move of the climatic zones, but fragmented habitats will make changes in range more difficult than in the past. A few may even manage to evolve swiftly enough to adapt to the new conditions. But many animals and plants that today flourish so abundantly, precisely because they are so well adapted to the conditions in which they live, will be unable to respond and as a consequence will become extinct.

What is to be done to prevent the Earth from becoming so swiftly and drastically impoverished? The problems involved are of great complexity. To tackle them, human beings will need all their knowledge of the workings of the living world. For that reason alone, study of the biological sciences is more important to humans than ever before.

Ecological crisis—a global problem

And the problems exist also on a global scale. Discharging poisonous industrial gases into midwestern skies affects trees and lakes in Canada. Cutting down the rain forest in Brazil affects the rainfall across the entire continent of South America. Felling coniferous trees for firewood in Nepal, on the flanks of the Himalayas, causes devastating floods at the mouth of the Ganges in Bangladesh. So no nation can, by itself, solve the environmental problems that are likely to afflict it.

Environmental catastrophe is not just a vague threat. In parts of Africa it has already arrived and brought starvation and death to millions. Its looming shadow elsewhere has now become so threatening that nations all around the world are beginning to realize that they must cooperate to save themselves. So maybe the deepening ecological crisis will bring the unexpected boon of making the nations forget at last the quarrels and wars that were the greatest disasters of the past and lead them to join together to forestall the even greater disaster that could lie ahead.

Human origins

The first and most important step in human evolution away from the apes was quite literally a step, for our distinctiveness began with the development of a more upright posture and a bipedal stride. The evolutionary advantage in that change, which took place some five million years ago, could well have been the freeing of the hands for carrying food back to the family or tribal group. So from the very start, the evolution of our species may have been associated with our social habits.

The role of climatic change

There is now little doubt that humans first appeared in Africa, and much of our early evolution occurred in response to the climatic changes that took place there between ten million and four million years ago. The great Antarctic ice cap expanded, and the climate of eastern and central Africa became drier. The forests in which apes had flourished contracted and were replaced by grasslands. These supported a great variety of bovid herbivores, such as antelopes, wildebeests, and gazelles, and many carnivores such as leopards, cheetahs, lions, hyenas, and hunting dogs.

This was the environment in which our ancestors had to survive. They were probably omnivorous, gathering vegetable foods—leaves, seeds, fruits, berries, and roots—together with insects, grubs, and the like. They probably also ate large mammals, but we cannot tell whether they obtained meat by hunting or by scavenging from the carcasses left by the carnivores.

The first humans

The earliest known member of our lineage is *Australopithecus ramidus*, who lived in East Africa four to five million years ago. Over the next three million years, two types began to diverge from this ancestral stock but continued to live side by side. One line, including *Paranthropus robustus*, developed heavy muscular jaws and powerful crushing teeth in order to eat tougher roots, tubers, and nuts. The other, which included *Paranthropus*

boisei, remained more lightly built and fed on a greater diversity of food. Contemporary deposits show that one or perhaps both types started to use primitive stone tools about two and a half million years ago.

Our own genus, *Homo*, evolved from another early hominoid *Australopithecus afarensis* a little over two million years ago. Taller and with a larger brain and more upright posture, *Homo erectus* coexisted with the australopithecines for some time before its predecessors became extinct.

These people made technological, and presumably social, progress. Their tools became more sophisticated and they learned to use fire, developments which made existing environments more productive and opened up new ones to habitation. Early humans began to spread out of Africa into cooler lands. By 700,000 years ago, *Homo erectus* people had traveled into Europe, to China and Southeast Asia, and even to the most northern parts of Eurasia.

Our own species, *Homo sapiens*, evolved in Africa from *Homo erectus* about 250,000 years ago. In fact, several distinct types of *Homo sapiens* seem to have succeeded one another. At least some of the earliest European fossils belong to a "Neanderthal" type, which was stockily built, large-brained, with a rather protruding nose and jaws and heavy brow ridges above the eyes.

But for the origins of modern *Homo sapiens* we must look once again to Africa. The oldest fossils of this kind are about 100,000 years old and have been found in eastern and southern Africa; the next spread of people out of Africa started about 10,000 years later. Studies of today's biochemical and genetic differences, such as those found in our blood groups and DNA, lend support to this picture.

To begin with, migration seems to have been slow. Life in these northern, cooler lands, with less game to hunt and a more seasonal supply of plant food, demanded the invention of a more complex array of tools. Of those that have survived—mostly the ones made of flint, bone, and

Australopithecus afarensis

ivory—many were clearly used for preparing plant food, for hunting and butchering animals, and for transforming their skins into clothing for warmth.

Colonizing the globe

It was as gatherers and hunters that these developing peoples colonized the globe and adapted to its climates. By 35,000 to 40,000 years ago, modern humans had spread throughout Eurasia and into Australia. It is likely that the New World was colonized from Asia shortly afterward, across the land bridge that then connected Siberia with Alaska. The last spots on Earth to be peopled were the isolated islands of the Pacific, which were reached only in the last few thousand years.

The next great chapters in human prehistory involved the tilling of the earth, the "agricultural revolutions" that seem to have taken place independently in different parts of the world from about 10,000 years ago onward.

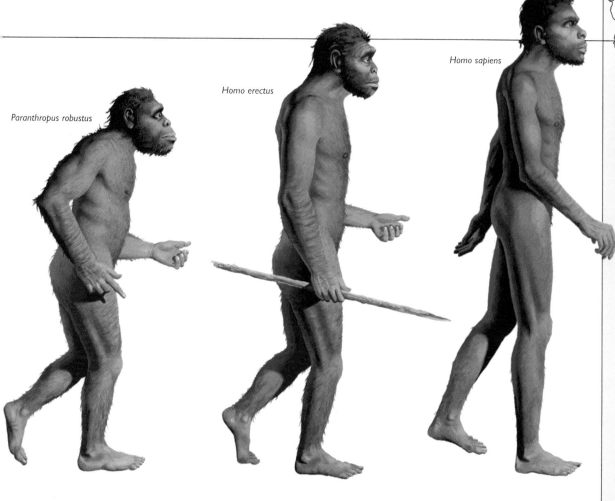

Paranthropus robustus

Homo erectus

Homo sapiens

When early humans moved into different environments away from Africa, they gradually adapted to new conditions. Northern peoples lost much of the skin pigment, which was a barrier to sunlight, thereby allowing the body to produce more vitamin D. Narrow eyes protected people from sun glare in desert or snowfield, and rounder, fat-padded faces gave better protection against the cold.

The spread of humans

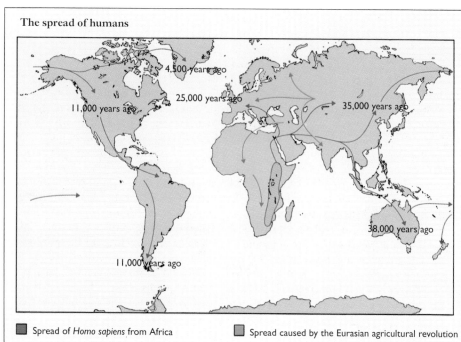

4,500 years ago

25,000 years ago

11,000 years ago

35,000 years ago

38,000 years ago

11,000 years ago

■ Spread of *Homo sapiens* from Africa   ■ Spread caused by the Eurasian agricultural revolution

The agricultural revolutions that began about 10,000 years ago led to further movements of human populations. One wave of expansion was that of farming peoples through Europe and India. Another wave pushed into northern Africa as far as the equator, forcing the original inhabitants southward. The latter became what are now known as the Bantu of southern Africa, whose own agricultural revolution took place only about 2,000 years ago.

Origins of agriculture

Hunting wild animals and gathering vegetable materials from their surroundings provided an adequate source of food and clothing for human populations until the beginning of the present interglacial 10,000 years ago.

Populations then began to grow and people discovered the advantages of managing both habitats and wild plant and animal species. This increased productivity and thus eased the burden and risks of hunting and gathering. A relationship developed between human beings and particular wild species of plants and animals that has become almost symbiotic—a mutual dependence.

Some animals, such as the wolf (dog), shared the hunting habit and were domesticated early, during the last glacial period. At about this time the productive steppe grasslands of the Middle East, Southeast Asia, and Central America supported game animals and contained plant species suitable for human exploitation. The development of arable agriculture started in these locations, with the collection of wild seeds in open, mixed vegetation. This led to the storage and sowing of seeds in controlled conditions where water could be supplied and harvesting was made easier.

Many of the early domesticated plants were annual grasses, such as wheat, maize, and rice; these plants invest much of their growth effort into the production of seed. Annuals were well suited to the open grassland areas in which agriculture was first pursued, where the habitat was unstable and subject to catastrophe such as drought, fire, and trampling by grazing wild animals. Under such conditions an opportunistic plant, such as an annual grass, can thrive. Its optimal strategy is a short lifespan, rapid growth and maturity, and the production of as much seed as possible. The rich store of carbohydrates and proteins in the seeds made them very attractive to the humans who hunted game on these grasslands.

No doubt they discovered that these annual, opportunistic plants were most successful around their settlements and in areas where they had used fire for driving game. Habitat management by fire would have proved valuable both in the improvement of the grassland for grazing and in the encouragement of seed-producing annuals. From this it was a short step to the investment of some of the gathered seed in the sowing of a crop for the following year.

Once the idea of maintaining a species under close control had developed, the adoption of new species would have been simple. Rye probably first became known to agricultural humans as a weed species of wheat crops. Some of the early domesticated animals, such as sheep and goats, may have been captured raiding the primitive cereal fields. Wheat, barley, and lentils were probably domesticated at least 9,000–11,000 years ago but possibly as long as 18,000 years ago. Rice was probably domesticated as long as 10,000–15,000 years ago.

Date of first evidence of domestication

12,000 years ago
11,000
10,000
9,000
8,000
7,000
6,000
5,000
4,000
3,000
2,000
1,000

Dog

Scarlet runner bean

Squash
Gourd

Common bean

Yam

Common bean
Lima bean

Sunflower
Tepary bean

Maize
Tomato
Sieva bean
Scarlet runner bean
Cotton
Avocado
Papaya
Cacao

Guinea pig
Llama
Alpaca

Pineapple

Potato
Peanut
Lima bean

The main centers of early domestication are shown on this map, together with the plant and animal species involved. When dates for the first domestication of particular animals and plants are known, these are indicated by color-coded flags. Where dates are not known, centers are indicated by poles without flags.

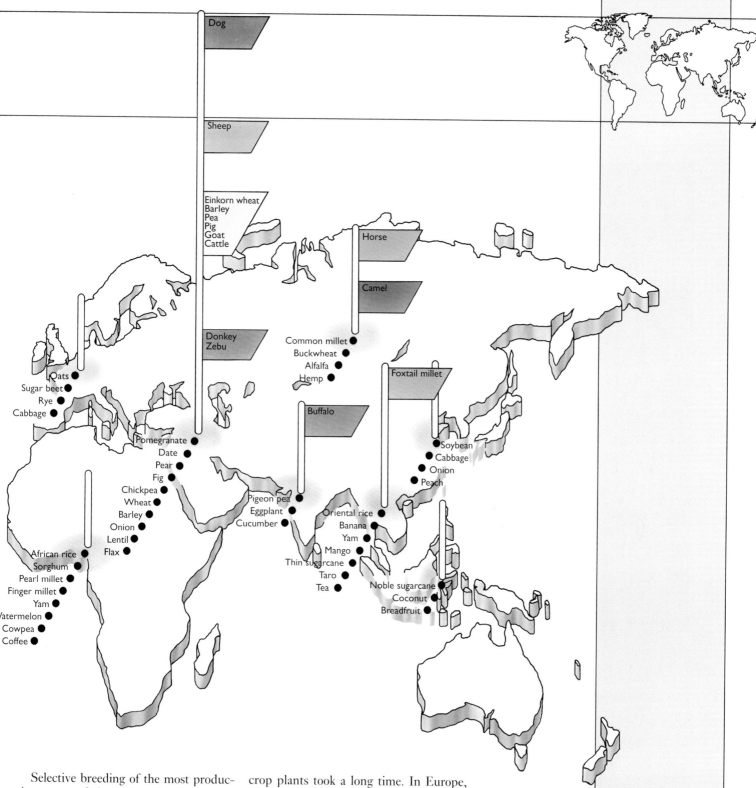

Dog

Sheep

Einkorn wheat
Barley
Pea
Pig
Goat
Cattle

Horse

Camel

Donkey
Zebu

Common millet ●
Buckwheat ●
Alfalfa ●
Hemp ●

Foxtail millet

● Oats
Sugar beet ●
Rye ●
Cabbage ●

Buffalo

● Soybean
● Cabbage
● Onion
● Peach

Pomegranate ●
Date ●
Pear ●
Fig ●
Chickpea ●
Wheat ●
Barley ●
Onion ●
Lentil ●
Flax ●

Pigeon pea ●
Eggplant ●
Cucumber ●

Oriental rice ●
Banana ●
Yam ●
Mango ●
Thin sugarcane ●
Taro ●
Tea ●

African rice ●
Sorghum ●
Pearl millet ●
Finger millet ●
Yam ●
Watermelon ●
Cowpea ●
Coffee ●

Noble sugarcane ●
Coconut ●
Breadfruit ●

Selective breeding of the most productive types of domesticated plants came under human control; those individuals most suitable for cultivation were favored and perpetuated. In annual species, such as cereals, squash, peas, beans, and lentils, selective breeding could proceed quite quickly and crop improvements would have been rapid.

The spread of the agricultural idea and of the raw materials in the form of seeds of crop plants took a long time. In Europe, where it is well documented, the expansion of farming beyond the Mediterranean region may have been limited by the need to clear heavy forest farther north in order to grow crops. Barley, for example, was cultivated in the Middle East more than 9,000 years ago. It reached southeast Europe around 7,000, northwest Europe about 6,000 and Britain 5,000 years ago—hardly a rapid spread.

The story
of wheat

One of the most familiar crops in temperate areas of the world is wheat, the descendant of wild grasses from the Middle East. First brought into cultivation many thousands of years ago, bread wheat (*Triticum aestivum*) has played an important role in the development of human civilization. But the story of its origins is only just emerging.

Exactly when wheat was first cultivated, rather than just gathered from the wild, is difficult to determine. Some of the oldest archaeological finds of wheat grains are from Egypt and date from about 18,000 years ago. The location of these finds, far south in the Nile Valley, is well outside the probable wild range of the original species, and the remains are therefore likely to be domestic.

There are about 20 species of wild grasses of the genus *Triticum*, all found in western Asia and around the eastern edge of the Mediterranean. They can be divided into groups based on the type of genetic material contained in their cells. Each cell of a plant or animal contains genetic material (DNA) in its chromosomes (see pp. 16–17). The precise number of chromosomes varies from species to species.

In a mature plant cell there are two sets of chromosomes (2N), one set from each parent. When pollen grains or egg cells in the ovary are formed, the two sets separate and each pollen grain or egg contains only one set (N). When the two reproductive structures fuse together after pollination and fertilization, they make a cell with two sets of chromosomes (2N) from which the seed develops.

The number of chromosomes in wheat cells provides invaluable clues to the way in which this staple crop has developed. And from species to species, the number of chromosomes differs. The smallest number found is 14 (seven pairs) in einkorn (*Triticum monococcum*) and some other wild species. Another group of wheats, which includes emmer (*T. turgidum*), has 28 chromosomes. The modern cultivated wheat (*T. aestivum*) has 42 chromosomes, three times the number of the primitive species.

Precisely how these groups of wheats are related is as yet unclear, but geneticists have produced a likely scheme of evolution (right). Einkorn and other primitive wheats with a low chromosome number were the first of the wild wheats. If two of them, such as einkorn and

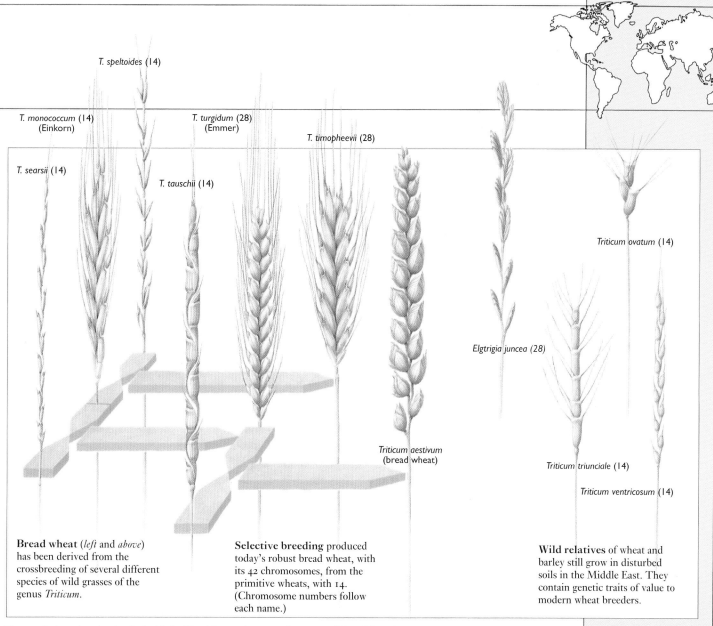

T. speltoides (14)

T. monococcum (14)
(Einkorn)

T. turgidum (28)
(Emmer)

T. timopheevii (28)

T. searsii (14)

T. tauschii (14)

Triticum ovatum (14)

Elgtrigia juncea (28)

Triticum aestivum
(bread wheat)

Triticum triunciale (14)

Triticum ventricosum (14)

Bread wheat (*left* and *above*) has been derived from the crossbreeding of several different species of wild grasses of the genus *Triticum*.

Selective breeding produced today's robust bread wheat, with its 42 chromosomes, from the primitive wheats, with 14. (Chromosome numbers follow each name.)

Wild relatives of wheat and barley still grow in disturbed soils in the Middle East. They contain genetic traits of value to modern wheat breeders.

T. searsii, interbred, the hybrid (still with 14 chromosomes) would have been able to grow but not reproduce. Because their parents were too different, the chromosomes would not behave correctly during the production of pollen and egg cells, and the hybrid would have been sterile.

What happened next was an evolutionary event that overcame the sterility problem. The number of chromosomes in each cell doubled, so that there were two sets from each parent instead of one. Pollen and eggs could form successfully, but each reproductive cell now contained 14, rather than 7, chromosomes. The adult hybrid, formed as a result of a 14-chromosome pollen cell and a 14-chromosome egg cell, had 28 (4N).

The fertile hybrid wheat, emmer (*T. turgidum*), with 28 chromosomes, was

robust and successful, with a better yield of grain than the primitive wheat. It was spread throughout Europe in prehistoric times and is still cultivated in parts of the Middle East. But in Iran, some 8,000 years ago, the next step in wheat evolution occurred. There the 28-chromosome emmer wheat met up with another wild wheat species (*T. tauschii*), with 14 chromosomes like emmer's parents. The emmer interbred with the wild species, and another sterile hybrid resulted, with 21 chromosomes (14 from emmer and 7 from *T. tauschii*). Once again the sterility problem was overcome by an increase in chromosome numbers. The chromosome number doubled to 42 to produce the fertile hybrid bread wheat (*T. aestivum*), highly productive and full of vigor and variety, that is used today.

About 20,000 cultivated varieties of bread wheat now exist as a result of selective breeding.

Yet there are still many ways of improving strains to help them cope with certain conditions. Conservation of wild wheats is vital. The genes they have developed in response to pressure in the wild—grazing, pests, disease, and drought—are invaluable in helping modify and strengthen modern wheat.

Domesticated animals

As humans moved out of the warm, tropical lands where they had evolved into cooler, more temperate regions, they found fewer animals to hunt. And as people became more numerous, and more proficient hunters, this scarcity must have become more significant. The ability to rear and keep animals soon became an important factor in survival; and the tribes that acquired the skill to do this must have gained a marked advantage over their less well-provisioned neighbors.

Selective breeding

Once animals were domesticated, human control over them extended beyond their use for meat and milk, wool and hides, to their breeding. And by selective breeding humans were eventually able to manipulate an animal's physical characteristics to emphasize those that they could best exploit. This was possible because factors such as size, thickness of coat, size of jaw, teeth, or horns, and type of meat or fat are all under genetic control. Indeed, this way of producing totally different breeds of an animal was part of the evidence for evolution by natural selection that Charles Darwin set out in his classic book *The Origin of Species*.

Domestication probably began with the rearing as pets of the small young of adults that had been hunted and killed. And since humans began as hunters, it is not surprising that the first animal to be tamed was the dog. By rearing wolf cubs and establishing themselves as "pack leaders," as it were, people domesticated them as companion animals. By favoring those that barked readily, they could also use them to warn of the approach of predators or enemies. The earliest trace of a domesticated animal is a dog's jawbone about 12,000 years old, found in a cave in Iraq.

The earliest evidence of the cat's association with humans is in Egyptian tomb paintings of 3,500 years ago. Cats have long been domestic pets, providing pleasure and companionship; their useful role in a domestic setting has, however, been largely limited to controlling pest populations of mice and rats.

Domestication on any scale requires that animals have a social pattern of behavior so they can be raised in flocks and herds. They must also be reasonably placid, so that they can be controlled and confined without danger to the herdsman. But permanent systems for keeping and breeding domestic animals for food became practicable only when people became settled agriculturists. Evidence suggests that this occurred some 13,000 years ago in the Fertile Crescent of the Middle East, where there was an overlap in the natural ranges of the earliest livestock: sheep, goats, cattle, and pigs. The quantity of bones found near such early settlements indicates that these animals were a reliable source of food.

By about 8,500 years ago, several mammals of Eurasia had been domesticated for their meat or milk. Today's sheep and goats are probably descended from the Asiatic mouflon sheep and bezoar goat; cattle from the aurochs, or wild ox, that roamed through southern Europe and Asia; and the useful scavenging pig from wild boar living in the forests. Even horses and donkeys may originally have been kept for milk or meat, but it is likely that their usefulness in pulling carts and carrying loads soon became apparent.

In North America, people were less fortunate. The horse became extinct there about 8,000 years ago, and the only large grazing mammal was the bison—which has never been tamed. But in South America, the mountains were inhabited by the guanaco, a fast, agile, thick-coated relative of the camel. Domesticated about 4,500 years ago and used for food, wool, and as a draft animal, its descendants are the llama and alpaca.

Today the descendants of the first domesticated animals range widely throughout the world, though climate and parasites still make it difficult to breed them in tropical Africa. Many indigenous herbivores are, however, successful there. These include the buffalo, oryx, and eland as well as relatives of the pig such as the warthog, bushpig, and forest hog. Perhaps these animals, already adapted to life in the African tropics may in the future provide domesticated breeds to help supply food for people on that continent.

Domestic breeds
Dogs are thought to be the first animals to have been tamed and domesticated. The earliest evidence dates back some 12,000 years, when tamed wolf cubs were probably used as companions and to warn of approaching danger.

By breeding dogs to meet particular needs, humans subsequently developed different types for hunting, herding, and protection. Today more than 200 breeds are recognized.

Beagle

Shetland sheepdog

Dobermann pinscher

Cairn terrier

Golden retriever

A heavily laden horse struggles across a glacier in the Karakoram mountain range in Central Asia. For at least 5,000 years horses and mules have been used to pull or carry burdens—including humans.

Feeding
the world

Humans are just one of the many millions of animal species inhabiting the surface of the Earth. Like all the others, we are ultimately dependent on how much sunlight can be trapped and converted into energy by plants, and how much of that energy can be diverted into the food chains on which we depend for survival.

The whole globe is one massive ecosystem, and it can only maintain a finite number of people. Of the total terrestrial productivity from natural and agricultural ecosystems, human beings at present divert 5.5 percent directly for their use. This may be as food, forage for domestic animals, for energy used in cooking, or in timber for construction.

At the same time, however, we waste as much as 34.6 percent of the Earth's production. A large proportion of forest biomass is never harvested; large areas of pastoral grassland are never grazed by domestic animals, and much of our arable produce is never consumed. Another 5.4 percent of production is lost to the world through land degradation by urbanization, dereliction, and desertification. Thus almost a half (45.5 percent) of the Earth's terrestrial primary production is diverted by one species—our own.

Energy: input and output

The supply of food energy to humanity is far from evenly distributed through the world. Food is overproduced in some areas (mainly temperate) and under-

produced in others (mainly tropical). In Africa, India, and Pakistan, for example, food consumption per person per day represents 60 percent of the calorie intake and only 50 percent of the protein intake of people in North America and Western Europe.

In modern intensive farming, much energy is expended on growing a crop. Machinery is fueled by gasoline and is needed for plowing, planting, spreading fertilizers and pesticides, and for irrigating

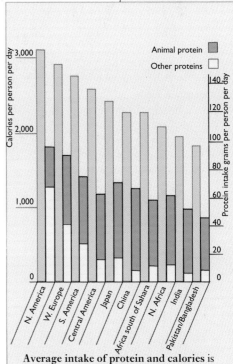

This is one of many experimental growing methods being tested in attempts to increase food production. Here, in the environmental research laboratory of the University of Arizona, lettuces can be produced without sun, rain, or soil. They are growing in a tunnel that rotates around an inner source of light

and harvesting. Fertilizers and pesticides demand much industrial energy in their manufacture. Transportation, preparation, and packaging of food also need fuel. For potatoes the amount of energy expended in these ways is only slightly less than that obtained in the food. Potato growing can be seen as a method of converting fossil fuel to starch.

And if the input of energy into animal products is compared with the output, the ratio is far worse: to produce one egg, for example, costs five times as much as its own energy value in fossil fuels.

In the United States, about 10 times more energy (mainly in the form of fossil fuels) is expended in agriculture than is actually provided by that agriculture. By contrast, in "less advanced" cultures, such as the hunter-gatherers, the energy invested in food production is only about one-fifth to one-tenth that extracted. Primitive agriculture, such as wet rice production in Southeast Asia, is even more energy efficient, with up to 50 times more extracted than expended.

But low-intensity, low-productivity agriculture can only be economical in areas where labor is cheap and there are fewer people to support per unit area of land cultivated. Increasing populations in the developing world have made improving agricultural production one of humankind's most pressing problems.

Biotechnology

Hope for the future comes from two main areas of research. The first involves bringing new species—many of them as yet unexploited—into cultivation. The second, biotechnology, will speed up breeding processes and furnish new crop varieties with novel genetic constitutions able to withstand the environmental difficulties of the developing world.

The wild genes of ancestral plants play an essential role in these developments—for we cannot make genes, only shuffle them. In fact, the unique genetic packages contained in wild plants and animals must be one of the strongest arguments for biological conservation.

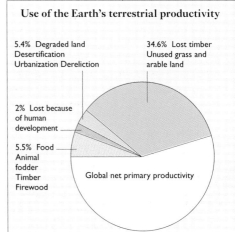

Use of the Earth's terrestrial productivity

5.4% Degraded land Desertification Urbanization Dereliction

34.6% Lost timber Unused grass and arable land

2% Lost because of human development

5.5% Food Animal fodder Timber Firewood

Global net primary productivity

Total: 45.5% diverted for human use

A total of 45.5 percent of the world's plant production on land is used by humans or wasted by human activity. Over 5 percent of potential production is lost either because that land has been destroyed by urbanization or because mismanagement has led to desertification.

Average intake of protein and calories is shown for different regions of the world. Consumption is most areas falls far short of that in the United States and Europe.

The spread
of weeds

A weed is essentially a plant pest, and the term may be used to describe any plant growing where it is not wanted. It may be an herb such as groundsel or ragwort in an arable field, an alien tree in a managed forest, moss in a garden, or an aquatic plant blocking a waterway.

The successful weed has a variety of attributes. Fast reproduction is usually necessary so that populations can be increased as quickly as possible. Thus most weeds, particularly annual species, invest much effort in seed production. The pineapple weed (*Matricaria discoidea*) can produce as many as 450 seeds on each flowering head.

Other weeds achieve rapid proliferation by vegetative means. Duckweed (*Lemna minor*) consists of small, disklike plates floating on water. Each of these can give rise to daughter disks, so that a pond or waterway soon becomes covered with the weed.

Another successful weed strategy is fast growth. Couch grass (*Elytrigia repens*), a persistent weed of agricultural ground, spreads rapidly by underground stems or rhizomes. If these are broken up by plowing or digging, each fragment can grow into a new plant, spreading the weed faster.

Good dispersal is a valuable feature in a weed, allowing it to invade a habitat ahead of other plants and gain an advantage. This may be achieved by seeds or by plant fragments that can grow into new plants.

Many weeds, such as poppies, have seeds with long viability in the soil; these may keep a bank of dormant seeds that can germinate rapidly when conditions are right.

The ideal weed must be relatively insensitive to the conditions of its environment, for such tolerance will give it the greatest flexibility when invading. The only situation unfavorable to most weeds is one of heavy shade, so they tend to establish themselves in less shaded, disturbed conditions, such as those produced by human activity.

The species that are the weeds of today grew on the planet long before humans arrived, and thrived in naturally open, disturbed habitats such as sand dunes, sea cliffs, river banks, and volcanic slopes.

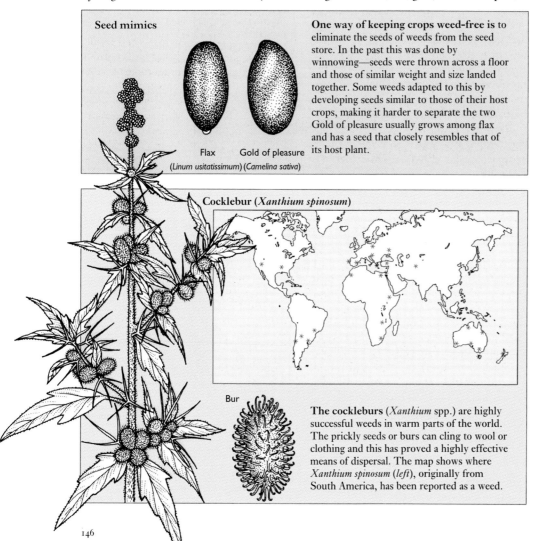

Seed mimics

Flax
(Linum usitatissimum)

Gold of pleasure
(Camelina sativa)

One way of keeping crops weed-free is to eliminate the seeds of weeds from the seed store. In the past this was done by winnowing—seeds were thrown across a floor and those of similar weight and size landed together. Some weeds adapted to this by developing seeds similar to those of their host crops, making it harder to separate the two. Gold of pleasure usually grows among flax and has a seed that closely resembles that of its host plant.

Cocklebur (*Xanthium spinosum*)

Bur

The cockleburs (*Xanthium* spp.) are highly successful weeds in warm parts of the world. The prickly seeds or burs can cling to wool or clothing and this has proved a highly effective means of dispersal. The map shows where *Xanthium spinosum* (*left*), originally from South America, has been reported as a weed.

When humans first cleared areas of vegetation for agriculture, such open-habitat plants took advantage of the new, favorable conditions.

Traveling weeds

Many weeds have thrived particularly when transported far away from their native homes by man. Saint-John's wort (*Hypericum perforatum*) has been a serious pest in the prairies of North America but is not a problem in its native Europe. The prickly pear (*Opuntia ficus-indica*), a New World cactus, is a successful weed of hot, dry climates in the Old World, notably Australia, where it was eventually controlled by using an Argentinian moth that tunnels into its stems. North America has

supplied a wealth of weeds to the Old World—60 percent of the weed flora in Europe and 32 percent of that in Africa originated there.

In agricultural situations, a trick of successful weeds is to mimic crop plants with which they grow in close association so that they are harvested with the crop. Often such weeds include wild ancestors of the crop species in question. The wild oat, for example, is a serious pest of cereal crops because even seed screening cannot differentiate between crop plant and weed.

With increased human mobility, weeds have also found it easier to travel. The plantains, cockleburs, and even ferns, such as the water fern *Salvinia*, have spread around the world in the wake of humans.

A huge raft of water hyacinths in Malawi shows how this tropical plant can proliferate to such a degree that it blocks waterways and rivers. In India and Africa particularly, the water hyacinth (Eichornia crassipes) has become a serious pest.

A common roadside weed in North America, the pineapple weed (*Matricaria matricarioides*), probably first arrived there from northeast Asia. Its seeds adhere to mud and hence to car tires. So the plant spread quickly and efficiently with the expansion of the road system.

From North America it invaded Europe early in the 20th century. There, too, it spread dramatically with the increase in road traffic.

Some of the most successful weeds are unwittingly dispersed by humans.

The thanet cress (*Cardaria draba*), for example, arrived in Britain from the Mediterranean region in 1809. It was brought in the straw bedding of injured soldiers returning from Spain during the Napoleonic Wars. When a local farmer plowed the bedding into his fields as manure, the plant became firmly established.

Animal pests

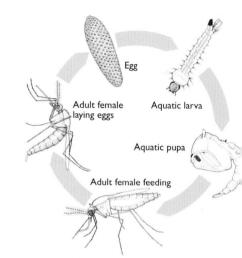

Red-billed queleas (*Quelea quelea*) nest and fly together in vast flocks, often more than a million strong. They can completely destroy large areas of crops.

Adult female mosquitoes lay eggs in freshwater. From these eggs hatch aquatic wingless larvae that feed on bacteria. The larvae pupate and transform into winged adults. These then leave the water and feed by sucking the blood of vertebrate hosts.

Life cycle of the mosquito

Egg

Adult female laying eggs

Aquatic larva

Aquatic pupa

Adult female feeding

An animal pest is an animal present in large numbers in the wrong place at the wrong time, and considered harmful, damaging, or annoying by the humans who share its habitat. Any type of animal can be a pest; the trouble they cause can range from the minor irritation of a flock of sparrows picking off crocus buds in a garden, to the life-threatening bites of malaria-infected mosquitoes.

Animal pests fall into four main categories. First are those that can harm humans or their farmed animals. Venomous snakes, spiders, and insects come into this group, as do the biting flies, bot-flies, screw worm flies, midges, and mosquitoes, whose bites are irritating, painful, or can cause skin damage. Second are pests that destroy or damage crops by consuming them or by reducing their growth. These include eelworm plant nematodes, caterpillars, locusts, plant bugs, hoppers, and aphids, as well as birds such as pigeons and queleas.

Stored product pests

A third and extremely important group has a great economic impact in all parts of the world. These are the stored product pests that consume stored human food. The main culprits are insects and small rodents, and their depredations may be on only a domestic scale—one mouse in a larder—or may occur on an industrial level with, for example, massive infestations of weevils in grain silos.

The final group do their damage principally by transmitting disease. Nearly all the vectors, as they are called, are insects, and they infect humans and animals with diseases such as malaria, yellow fever, plague, and sleeping sickness.

Almost all serious animal pests, like most plant pests (or weeds), have a high reproductive potential. And this ability to multiply rapidly gives them the power to cause harm at significant levels; it also makes controlling them more difficult.

Many of the most serious pests are insects. They are the most diverse animal group—there are millions of different species—and they can adapt to almost any conceivable niche. More than anything, though, it is their facility for rapid population expansion that makes insect pests such awesome foes, for a female may produce a new generation within a few days. Thus, insects can respond terrifyingly fast to new nutritional opportunities. Moreover, any control measures must have an extermination rate of close to 100 of percent if they are to have any chance of success.

The smaller an animal, the shorter its generation span. Insects are small, so it does not take them long to grow and reach reproductive maturity, which they do through a series of larval development stages. In some insects, such as the mosquito, these larval stages may not resemble the adult at all, while the nymphal larvae of others, such as the cockroach, all look very much like the adult. Whatever the larval development type, large numbers of offspring are produced with extreme rapidity.

A bird pest

As a rule, other noninsect types of stored product and crop-devouring pests also have high reproductive rates. Mice and rats live in this way and are among the most rapidly breeding mammals; the quelea is a remarkable example of a pest bird with this characteristic.

An African bird, about 5 in long, the red-billed quelea, or dioch, is related to the weaver birds. From earliest times it has been recognized as a devastating pest of grain crops such as millet, sorghum, and wheat. Its breeding rate is extremely high, with many broods in a season, and it is also highly gregarious—both factors that contribute to its pest status. When the flocks are on the wing, they have been likened to swarms of migrating locusts.

A grain weevil (*Sitophilus granarius*), seen through a scanning electron microscope, emerges from a wheat grain.

Stored product pests

Grain, flour, and other foodstuffs stored by humans provide a plentiful source of nourishment for a wide range of animals. These are known collectively as stored product pests, and while the term includes rats and mice, most are insects. They attack a variety of foods the domestic booklouse alone, for example, infests 50 different types, as well as books, paper, and other nonfood substances.

House mouse (*Mus musculus*)

Flour moth (*Ephestia kuehniella*)

Oriental cockroach (*Blatta orientalis*)

Booklouse (*Liposcelis*)

149

Controlling pests

Almost 3 million people die each year from malarial infections, all spread by mosquitoes. Often 10–30 percent of a food crop, or even more, will be lost to pest insects in the field and to stored product pests after harvest. The magnitude of these human and commercial losses has stimulated far-ranging research efforts to find ever better ways of controlling pests.

The diversity of approaches to control is great, but almost all fall into one of three categories—chemical, environmental, or biological. The technology of effective pest control is, however, fraught with difficulties, inevitable when the target is resilient and adaptable and breeds rapidly.

A parasitic disease

Both the diversity and the difficulties inherent in such an enterprise are clearly demonstrated by the attempt to eradicate the disease known as river blindness, or onchocerciasis. This is caused by the nematode *Onchocerca*, a tiny roundworm. A high proportion of the 15 million sufferers, who carry the worms in their bodies, endure persistently itching skin and progressive eye damage, leading to blindness, often by the age of 30. The disease is passed on through an intermediary, the blood-feeding, biting blackfly, *Simulium*. From the blood of an infected person, it takes larval parasitic worms into its own body, where they rapidly develop into new larvae. The next time the blackfly bites and feeds, another person is infected.

The distribution of the disease is tied to the distribution of the blackfly host, found in a band stretching across Africa, in the Yemen, and in Central America. In all these places, close to the rivers where the blackflies breed, the disease is present in human populations.

In West Africa, a massive and sustained effort—the Onchocerciasis Control Program, or OCP—is being made to eradicate the blackfly. It was started in 1974 and continues to the present day; funding has been provided by the World Bank, the World Health Organization, and the Food and Agricultural Organization. They aim to curtail blackfly breeding in all the rivers

River blindness and the blackfly

□ Control zone
□ Distribution of blackfly
□ Incidence of onchocerciasis (river blindness)

Blackflies (*Simulium*) are small biting insects that feed on human blood. In parts of Africa, as they feed they may transmit from person to person the nematode worms

Larva

that cause river blindness. Blackfly eggs are laid in rivers, and the developing larvae attach themselves to stones or vegetation under the water.

Blackfly

The distributions of the disease known as river blindness and the river-breeding blackfly that transmits it largely coincide, as seen here in Africa.

In part of West Africa a major control scheme has been in operation since 1974. Rivers in the area have been sprayed with chemical insecticide to kill the blackfly larvae.

in an area covering about 400,000 sq miles in eleven countries including Benin, Burkina Faso, Ivory Coast, Ghana, Mali, Niger, and Togo. More than 12,400 miles of rivers are at present covered by the program, which employs a variety of strategies.

The flies have been attacked by spraying a highly specific organophosphorus insecticide, temephos, in carefully monitored amounts into the rivers. The spraying is done from the air as well as by ground teams. Since the program began, blackfly breeding has almost ceased in the central parts of the control zone. This has stopped all new onchocerciasis infections in children in that area.

The program, however, ran into some difficulties. Blackflies can enter the control area from outside, but this "edge effect"

problem has been countered since 1986 by extending the scheme into eastern Mali, Guinea Bissau, Senegal, and Sierra Leone. Another problem arose when some local blackfly populations developed a resistance to temephos; this necessitated a switch to a different insecticide.

An unorthodox biological control agent has also been used. This consists of the commercially produced spores of an insect-killing bacterium, which will, in theory, perpetuate itself by growing inside the flies and killing them. Environmental control measures, too, have been introduced: the design of spillways and canal and dam walls has been altered so that larvae cannot easily attach themselves to them. This disrupts the development and breeding pattern of the flies. A preventative drug, mectizan, is now available.

Several plant species produce chemicals that protect them from herbivorous insects. Some, for example, manufacture substances that mimic the hormones controlling insect molting and development.

Industrial chemists, using these "false hormones" as starting points, have produced insecticides that inhibit or disrupt pest insect development. Their great advantage is that they are entirely specific and have no harmful effects on other organisms.

The flowers of the chrysanthemum genus Tanacetum are the source of a group of natural insecticides (pyrethrum) that are among the most efficient known.

The screw worm

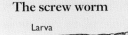
Larva

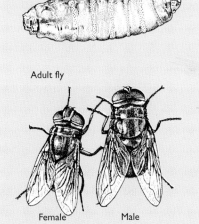
Adult fly

Female   Male

A subtle form of biological control has been used in the United States to combat the screw worm fly, a cattle pest. Female flies mate once only, then lay their eggs on the skin of a cow. The fertilized eggs contain the maggotlike larvae (screw worms) which, when they hatch, bore through the cow's hide to feed on its flesh. Here, the larvae pupate and eventually emerge as adult flies through the hide.

The "sterile male" control technique has been successfully used against the fly. Millions of males, bred on an industrial scale, are exposed to a dose of gamma radiation that renders their sperm sterile. When they are released, these sterile males mate with local females. The eggs are not fertilized and so do not hatch. This greatly reduces the fly's breeding success.

Traveling with humans

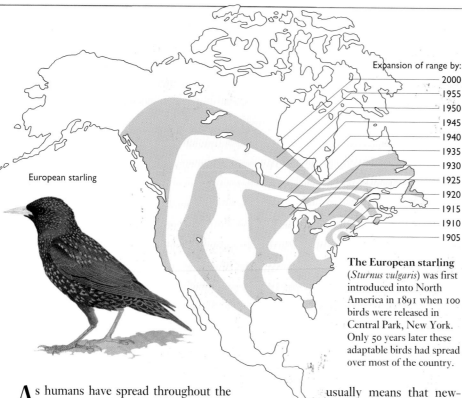

European starling

Expansion of range by:
2000
1955
1950
1945
1940
1935
1930
1925
1920
1915
1910
1905

The European starling
(*Sturnus vulgaris*) was first
introduced into North
America in 1891 when 100
birds were released in
Central Park, New York.
Only 50 years later these
adaptable birds had spread
over most of the country.

As humans have spread throughout the world they have carried animals and plants along with them. Sometimes this is done intentionally, to provide food, fur, sport, or visual delight. Often it is accidental—rats and mice, for example, have spread all over the world as unseen passengers on ships.

The introduction of new organisms into an environment is risky. The host ecosystem has evolved a natural balance in which the numbers of any animal or plant species are controlled by the abundance of its food, predators, parasites, and competitors. This usually means that new-comers are confronted by a highly efficient "closed shop." For example, of over 100 species of birds introduced into Britain, only four have established themselves over a wide area—the Asian pheasant, the redlegged partridge, the little owl, and the Canada goose.

Disturbing the balance

On the other hand, a species that can find a vacant niche within the host community or can displace an existing member of that community may find itself not only with abundant food but free of the predators and parasites of its normal range. Such fortunate intruders are then likely to run rampant, building up enormous numbers and upsetting the balance of their adoptive ecosystem. Imported species account for around 60 percent of agriculture's most serious pests and 20 percent of all mammal and bird extinctions over the last 400 years. Eventually a new balance would probably evolve, but the time needed is likely to be many thousands of years longer than the mere handful needed to destroy some species or to bring financial ruin to farmers.

The earliest known example of the transportation of an animal by humans is that of the dingo. This domesticated variety of the sheepdog of India was brought to Australia more than 3,000 years ago by aboriginal colonizers. It was probably responsible for the extinction of its native Australian equivalent, the thylacine wolf. Today, though it preys mainly on kangaroos, wallabies, and rats, the dingo also kills significant numbers of sheep and calves.

For this reason efforts have long been made to control the population by shooting or trapping animals, dropping poisoned meat from the air, and erecting a supposedly "dingo-proof" fence that stretched hundreds of miles across the country.

Australia and New Zealand, both of which remained relatively isolated for so long, have been particularly vulnerable to the introduction of new species. The European rabbit *Oryctolagus cuniculus*, for exam-

Prickly pear (*Opuntia sp.*)

Prickly pears (*Opuntia spp.*) were first introduced into Australia from the United States in 1839 as interesting garden plants. But conditions were so favorable that they spread rapidly and by the 1920s occupied some 50 million acres. One of the reasons for their success was that their normal predator—the caterpillar of the cinnabar moth—was absent. Once introduced, the caterpillars cleared the prickly pear from more than 95 percent of the affected area.

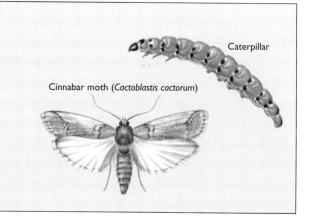

Caterpillar

Cinnabar moth (*Cactoblastis cactorum*)

ple, first arrived in Australia only 200 years ago, with the prickly pear cactus *Opuntia spp.* following shortly after. Both quickly achieved pest status.

In addition to rabbits, imports into New Zealand that have reached pest numbers include the European red deer (elk). Introduced in 1851 for sport and food, it now causes great damage to forests and competes with grazing livestock.

Two-way traffic

The opening up of North and South America to European travelers brought about a reciprocal "trade" involving thousands of species of animals and plants. The muskrat and nutria were introduced into Europe for farming of their pelts, but many escaped and established themselves in new wetland habitats.

Examples of transmission in the opposite direction include the European starling, whose invasion of North America has been mapped with great precision, and the brilliantly colored fireweed, now widespread in South America.

Some plants and animals have fared even better by their association with human travel. Spectacular examples include the now-ubiquitous European sparrow, at home in environments as diverse as Hawaii, the Falkland Islands, and northern European cities; and the South American water hyacinth (*Eichornia crassipes*), which has become a menace to navigation, fishing, and irrigation schemes throughout the tropics.

The dingo (*Canis dingo*) was first taken to Australia by humans some 3,000 years ago.

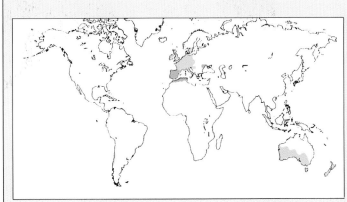

When European rabbits were brought to Australia, they spread so rapidly that they soon became the worst pest the country had ever known. In 1950 the virus disease myxomatosis was introduced; it killed 90 percent of the rabbit population in a short time, effectively bringing it under control.

◼ Natural distribution
◻ Naturalized distribution

City living

Bat
roosts in tops of buildings and attics; feeds on insects that cluster around street lighting

Parakeet
roosts in warm air ducts in North American cities

Brown rat
lives in sewers and basements and anywhere it can find food

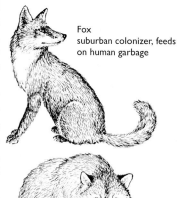

Fox
suburban colonizer, feeds on human garbage

Raccoon
suburban colonizer, feeds on human garbage

As the human population has grown to its present total of more than six billion, more and more of the Earth's surface (around 2 percent) has been covered by towns and cities. This built environment, known as the urban biome, must now be considered an identifiable habitat for plants and animals.

The animals that have taken advantage of this special habitat are an intriguing mixture of species: silverfish, mice, rats, squirrels, foxes, raccoons, opossums, and flocks of starlings, for example. They are mostly characterized by their ability to make use of a wide range of resources relating to food, shelter, and survival, and each animal has adapted to the city in its own particular way.

The disadvantages of city life are clear, including massive human interference, a relative lack of natural vegetation and soil surface, and pollution. The main advantages are less obvious—new shelters, new food sources, and new microclimates—but they are of great significance to an adaptable animal.

A city provides an abundance of shelter and nest sites. Houses have spaces under the floorboards, in cavity walls, and in the roof area; streets have underground cable ducting, pipes, and sewers; while roofs, walls, and windowsills provide what amounts to a cliff habitat full of crevices.

For some animals the use is casual; for others it has become their preferred way of living. Swifts nest on buildings, and house mice are found only in human habitations. This binding of animals to humans can become a defining characteristic. On the west coast of America, Brewer's blackbirds have taken up residence in shopping malls, feeding in restaurants.

Human populations store huge amounts of food and throw away large quantities of edible refuse; both sources are exploited by city-dwelling animals such as rats, mice, cockroaches, weevils, and flour beetles. Although regarded as pests, these animals are simply utilizing the superabundance of food provided by humans. Societies with food surpluses even provide food specifically for wild

animals: there can be few European and North American suburbs without birdfeeders. The extra, high-energy foods placed in them can have a considerable impact on the survival of city birds in harsh winter conditions.

Other new food resources available in cities have indirectly made the transfer from rural to urban living easy for animals such as the red fox. This intelligent, adaptable omnivore can now be sighted

close to the center of Paris, New York, Toronto, Amsterdam, Stockholm, and London. Lying low during the day in a den under a garden shed, for instance, and foraging in gardens and waste ground at night, it has adapted well to city life.

As well as providing new food sources and types of shelter, in cool temperate and cold climates the city protects animals from the winter weather. Some animals can find their way into warmed houses.

wl
n buildings

Pharaoh ant
lives under flooring in
large centrally heated
buildings such as hospitals

Blackbird
nests anywhere in cities
where there is a patch of
vegetation; tolerates
human disturbance

House sparrow
only nests in buildings;
opportunistic
feeder

Swift
nests in crannies in
buildings

Silverfish
primitive wingless insect;
feeds on detritus in
buildings

Furniture beetle
wood-boring beetle
adapted to feeding on
indoor wood

Flea
larvae of human and
domestic animal fleas
develop inside buildings

Housefly
worldwide insect pest;
feeds on any organic
material and spreads
bacterial contamination

House spider
traps other house insects
with its web

Others find a haven in city streets, which, owing to heat leaks from buildings, are significantly warmer than the surrounding countryside. In addition, the mass of buildings reduces wind speed and hence the wind chill factor. Indeed, the mass of brick and concrete acts like a huge storage device, warming up during the day and slowly releasing heat at night. This "heat island" aspect of a city may explain why the winter survival rate of warm-blooded mammals and birds is better there than in the country, and why city birds start to breed earlier than their counterparts in the country.

One telling set of statistics sums up the advantages of the city. After almost a month of exceedingly cold weather and snow cover in London in 1968, city black-birds (a type of thrush) weighted 5 oz on average, while those in woodlands outside London weighed only 3 oz.

A range of animal species, from racoons to fleas, has become adapted to an urban environment. The illustration shows a selection of such species and indicates some of the advantages they gain from life in towns and cities.

The spread of deserts

In only 70 years, between 1882 and 1952, the proportion of the Earth's land surface classified as desert rose from 9.4 to 23.3 percent. With desert still spreading today, it is vital to discover why the process is taking place—whether it is due to climate or is a consequence of human mismanagement.

The areas affected are mostly the semi-arid lands around the fringes of the great deserts. Plants can grow in semi-arid areas, and humans can make a living largely by keeping domestic grazing animals. But in the last few decades much of this land has suffered drought, and the vegetation has been degraded by overgrazing to such an extent that the land has become desert.

Even some of the older desert areas were not always so, and there is plenty of evidence of how conditions have changed in the long term. Many deserts, such as the Negev in Israel, have deep canyons running through them; their presence suggests there was formerly a considerable flow of water. Pollen analysis of the sediment from ancient lake sites in the Rajasthan Desert in India shows that some 6,000–9,000 years ago these were not only flourishing freshwater lakes, they were also surrounded by a relatively rich vegetation. There are similar indications in many other deserts that in the early part of the present interglacial their climate was far wetter than it is now.

Although meteorological records for deserts do not date back long, they indicate that there have been profound changes even in very recent times. Records for the last 70 years from the Sudan in Africa show that there has been a drop of about 15 percent in annual rainfall since 1920. The decline has been most dramatic since 1960 and is not getting better. Moving south of the Sahara, there is a sharp, progressive decrease of about 1 in for every 15 miles traveled. Such small shifts in rainfall pattern can have devastating effects. Here, drought spread south by about 5.5 miles a year during the 1970s and 80s. At the same time, winter rainfall in the Mediterranean has been increasing.

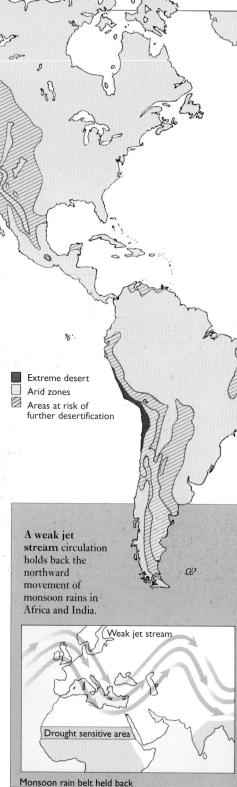

Changing climate and, in some areas, overgrazing are putting the arid and semi-arid lands on extreme desert fringes under stress. The map shows the large areas of the world in danger of becoming true desert if present conditions continue.

The reason for the current drought in Africa may lie in the strength of the wind patterns in temperate Europe. When the "jet stream" of air over Europe is weak it brings winter rain to the Mediterranean, but it prevents the penetration of monsoon winds and rain from the south into the desert regions of Africa and of northwest India; so drought results.

The human factor

Desertification is not only a consequence of climatic change. The human contribution has been considerable. Economic pressures cause herdsman to attempt to maintain large stocks of sheep and goats, even during drought. South of the Sahara political boundaries have prevented the free migration of herds with the changing rainfall and vegetation patterns, and this has exacerbated the overgrazing problem.

It is clear that the global climate is changing, resulting in poorer rainfall in many sensitive areas. Human pressures through domestic animals place an unsupportable strain upon the stressed vegetation; and as plant life is progressively eliminated, the physical conditions of the land deteriorate and desert becomes firmly established. Recovery is difficult but is being achieved by afforestation schemes in, for example, Israel and India.

Meanwhile, changes are taking place in the Earth's atmosphere that are leading to an increase in global temperatures (see pp. 166–67). These are likely to accelerate the course of desertification in the future.

Extreme desert
Arid zones
Areas at risk of further desertification

A weak jet stream circulation holds back the northward movement of monsoon rains in Africa and India.

Weak jet stream

Drought sensitive area

Monsoon rain belt held back

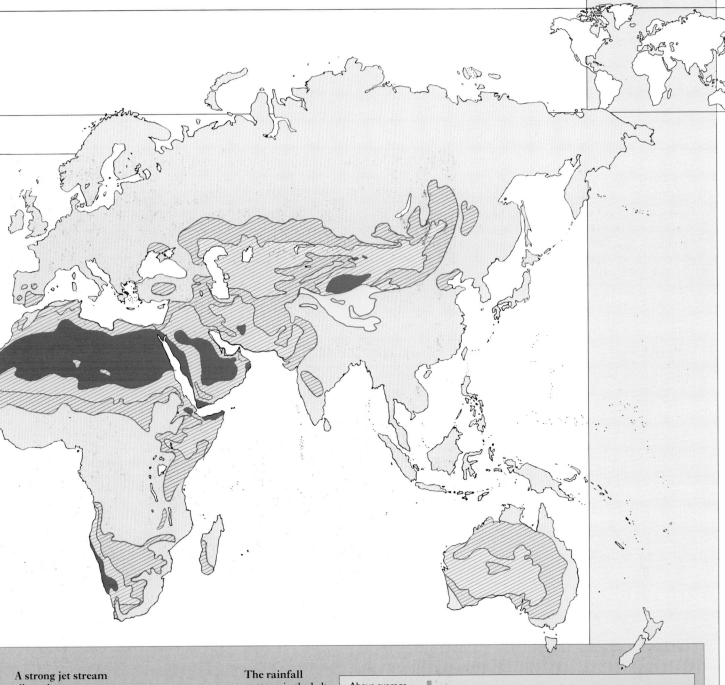

A strong jet stream allows the monsoon rains to penetrate the arid zones of Africa and India.

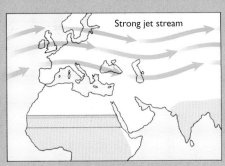

Monsoon rain penetrates drought-sensitive area

The rainfall pattern in the belt south of the Sahara Desert in Africa has changed considerably during the last half century. Before 1960 rainfall was generally above average, but since the mid-1960s it has been well below. The diagram shows departures from normal figures since 1940.

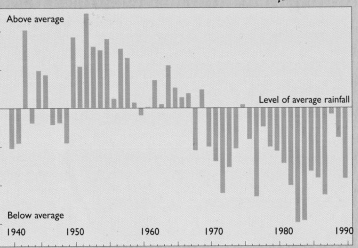

The ancient diseases

Tracing the patterns and distribution of the long list of infectious diseases humans are prey to may seem a morbid and negative exercise. The patterns of disease do, however, provide fascinating indirect clues about the origins, travels, and social development of humankind.

Although there is much overlap between the two, it is often convenient to divide human infectious diseases into ancient and recent categories. The ancient diseases are those that have been problems for hundreds of thousands, perhaps millions, of years; the newer diseases are those linked to more recent human history and development.

Lifestyle and disease

To find out about ancient diseases, it is necessary to understand the early evolution of our species (see pp. 136–37). The long-term diseases to inflict humankind have been with us since these distant and multiple origins. Those origins, and particularly the emergence of the modern subspecies of *Homo sapiens* about 100,000 years ago, all seem to have been rooted in Africa.

Our subspecies grew up in an evolutionary cradle that was entirely tropical, and that historical foundation of place and climate determined the pattern of many of our earliest diseases. These were, and remain, the infections to which mobile hunter-gatherers are susceptible: the bacterial diseases caught from shared waterholes, and the numerous diseases acquired by close ecological contact with near relative species and by the bites of blood-sucking insects and ticks.

Close relatives among species share diseases; this truism is certainly borne out by the pattern of early human disease. Most agents of infectious disease— viruses, bacteria, protozoans, and worms—are host specific, usually infecting only a few closely related species. And in the early days of our genus, two million years ago, we shared Africa with many monkeys and apes, our nearest cousins in the vertebrate family tree.

It is no surprise, therefore, to find that

today humans share the major infectious diseases such as yellow fever (caused by a virus) and some malarias (caused by protozoan parasites) with tropical apes. Both these diseases are spread from ape to ape, from person to person, or between two species, by the bites of particular tropical mosquitoes.

The primates, the group of vertebrates to which humans, the apes, and monkeys belong, are, almost exclusively among the major vertebrate groupings, restricted to the tropics. In its early days, humankind, like the other apes, was purely tropical in its distribution. The tropics, in turn, have

the greatest diversity and density of insects, so in two quite distinct ways, the risks of insect-transmitted infectious diseases have always been greater for humans in the tropics. First, there were many insects to spread the diseases; second, humans shared the tropical habitat with many closely related primate species from which new diseases were acquired.

Before each of humankind's expansions out of Africa, enough evidence was laid down to enable us to build a plausible picture of our niche there and the forces that held our numbers in check. Fossil and shelter finds in Africa show that early

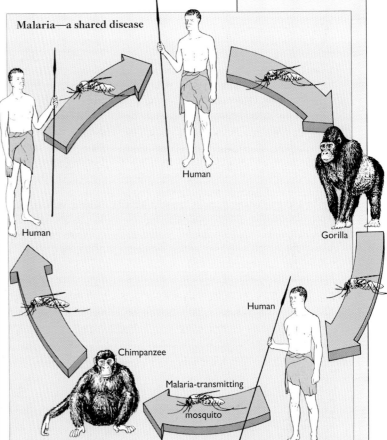

Malaria—a shared disease

Human

Human

Gorilla

Human

Chimpanzee

Malaria-transmitting
mosquito

Malaria is a tropical disease caused by blood-dwelling protozoan parasites of the genus *Plasmodium*. Different types of *Plasmodium*—all transmitted by mosquitoes—cause the disease in particular birds and mammals, and each type has a narrow range of vertebrate hosts. Some of the malaria strains that infect humans are shared by our close primate relatives, the gorilla and chimpanzee, reflecting an early period of our evolution when humans lived in Africa in close proximity to those other apes. The map (*below*) shows the world distribution of malaria.

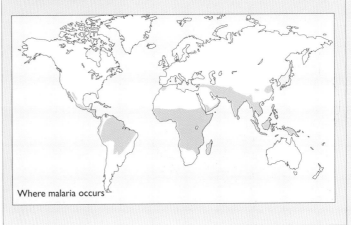

Where malaria occurs

people must have existed in a few small, family-based hunter-gatherer groups. Those early hominids were great nutritional opportunists, and in the productive tropics of Africa there were many foods to choose from.

It seems improbable, then, that early human populations in Africa were controlled by finite or restricted food resources. It is far more likely that the wide range of disabling, debilitating, and killing diseases of the African tropics were the single most important factor in keeping the human population density there so low for so long.

Fecal contamination of waterholes, such as this one in Kenya, can lead to cross-infections between species. Early humans in Africa probably became infected with many bacterial diseases in this way.

159

The new
diseases

With the movement of the "modern" subspecies of humankind out of Africa about 100,000 years ago, the most recent and fateful chapter of human history began: the peopling of the globe. This was the range expansion that was to provide all the geographically separated clusters of humans, who would ultimately evolve into the main racial groupings of today. The "break-out" from Africa also had other profound implications, some of them linked with disease patterns.

As the migrating groups moved north from the tropics, they encountered two fundamental environmental changes. First, they left all the other primates behind; in the temperate and cool zones of Europe, then Asia, and lastly the Americas, there were no monkeys and apes to share their habitat. No longer would they acquire infections from these closely related species.

Left behind, too, were the constant high densities of disease-transmitting insects that flourished in the warm climate of tropical Africa. There were fewer insects, and they largely disappeared in the cold winter months. These two influences meant that for the first time humans were not seriously impeded in their survival and reproduction by infectious diseases.

A new way of life

In the world's colder regions (colder than they are now because they were in the grip of a glacial), the major difficulty was to find adequate food and shelter. Caves and rock shelters, fires, animal skin coverings, and more sophisticated stone weapons and hunting techniques partly solved these problems. But living in caves and wearing clothes opened up the likelihood of new diseases. An infrequently removed animal skin covering provided a favorable protected habitat for skin ectoparasites such as body lice.

And in the caves where humans sheltered, new, specialized, blood-sucking insects found them tasty. One such unsought partner picked up by humans during this time was the bedbug. It belongs to a group of ectoparasitic insects that utilize cave-dwelling vertebrates exclusively; apart from humans, members of the family are found feeding only on caveroosting bats and birds.

By the end of the last glacial, around 10,000 years ago, all the habitable parts of the world, except for some remote oceanic islands, had been peopled. Then, as the

Floods, such as these in the Sudan, create ideal conditions for the spread of cholera in fecally contaminated water.

climate moderated, a series of linked changes occurred that were vastly to increase the population of *Homo sapiens*. These changes were the introduction of arable agriculture, the domestication of animals, metal smelting, and the growth of urban living.

In a good season, a person with a bronze or flint sickle could reap nearly 6.5 lb of wild wheat in an hour. A family could gather supplies for a whole winter during a three-week harvest period. The pattern of fixed settlements engendered by agriculture and the high reproductive success of their well-fed inhabitants meant that people were suddenly living close to large numbers of other people for the first time.

The rise of epidemic disease

High population densities, poor sanitation, and poor hygiene are conditions that predispose communities to epidemics. Bacterial and viral diseases, which persist with difficulty in small, mobile hunter-gatherer groups, could flourish in the villages and towns. It is likely that a range of diseases became widespread at this time; these would have included tuberculosis and influenza, since both need high-density human living to survive.

In the complex new world of settlements and agriculture, humans were also living in close proximity to their recently domesticated farm animals. Dogs, descended from wolves, were helping hunters to catch game as long as 12,000 years ago. Soon afterward, sheep, pigs, goats, and cows were under human protection for the sake of the food and skins they produced. They were followed by the transportation animals: camels, donkeys, horses, and llamas. And, just as with the apes, humans' closeness to these animals led to their contracting their diseases.

The longer the period of domestication, the more diseases humans share with the animals. It is not surprising, therefore, to find that we share most with our oldest companion, the dog. Indeed, the list runs to more than 50 illnesses, including toxocariasis, which causes blindness, and hydatid disease.

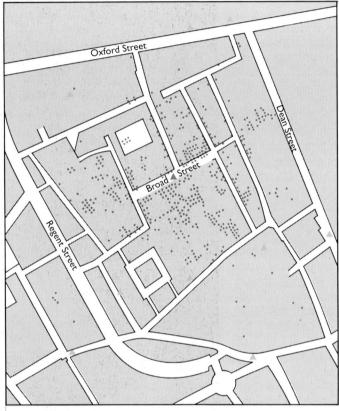

▲ Water pump
∷ Deaths from cholera

Cholera, an extremely severe form of dysentery, is spread by fecally contaminated water. Its chances of transmission are highest when large numbers of people are living in crowded conditions with poor sanitation.

In the 1850s such conditions existed in central London. John Snow, a physician at the time, was the first to realize that cholera was spread by water. In a detailed survey of deaths from cholera in the Soho area of London (see map, *left*), he revealed a cluster of cases around Broad Street. These were eventually linked to a public water pump, the source of which was contaminated by a fractured cesspit.

Contact between humans and animals can spread serious disease. The eggs of the parasitic roundworm *Toxocara* found in dogs and cats, for example, are easily transmitted to people. Children are particularly at risk, and in some cases infection causes eye damage or even blindness.

161

The spread
of disease

If it is understood how a disease is spread, it is often possible to think of a novel means of stemming its transmission. So the main goal of epidemiologists, who study the spread of diseases, is usually to find more effective ways of combating a disease.

Infective agents make the jump between human beings in one of six ways (see table). Some, such as the aquatic larval worms of bilharzia or those of the hookworm that live in the soil, bore into the human body through the skin. Simple precautions—avoiding skin contact with infected water and wearing shoes—can dramatically reduce such infestations.

Humans can also acquire diseases when they swallow food or drink contaminated by bacteria, bacterial spores, and parasite eggs. Many of these come from fecal contamination, a problem exacerbated by people living in crowded conditions with poor sanitation and little piped water. Improved living standards will often, on their own, reduce this type of infection. High population density is also responsible for the spread of many respiratory diseases by airborne droplets from coughs and sneezes.

The other modes of epidemiological spread are both biological in nature. The first is by vectors, other animals that carry the disease from one person to another. Insects are the main threat here: mosquitoes carry viruses such as yellow fever; rat fleas, the plague bacteria; blackflies, the nematodes causing river blindness.

Various preventive techniques, ranging from killing the animals to ingenious applied biological controls, are employed to combat these diseases (see pp. 150–51). Another potent protection is the use of immunization and vaccines, which boost the body's antibodies and cell-based immunity to a specific disease. In the case of the killer virus disease smallpox, global use of vaccine eliminated both the virus and the disease.

The second biological mode of spreading infection relates to the sexually transmitted diseases, such as gonorrhea and

syphilis, where the bacteria causing them pass directly between sexual partners.

On a broader scale, patterns of disease distribution and spread are less easy to categorize. Sometimes they are restricted by climatic variables. For instance, the snails that transmit bilharzia cannot tolerate cool temperatures or saline water; and yellow fever can only be contracted in a belt running through the equatorial tropics that is warm enough to support the breeding of *Aedes aegypti*, the mosquito that transmits it.

The disease of the 1980s
AIDS is caused by HIV, a retrovirus, a special type of virus that is, unusually, able to store its own genetic code in the form of RNA, rather than DNA (see pp. 16–17). Once it infects human cells, it can

produce a DNA version of its genes, then "hide" inside the human genetic code until it is later activated to produce new viruses. Several closely related forms of the virus seem to have evolved from similar retroviruses infecting monkeys, in which they seem to cause little or no disease.

The AIDS virus is normally passed on by the highly efficient sexual transmission route. But it can pass from mother to baby across the placenta or in breast milk, or into victims via contaminated blood or blood products in injection needles or by transfusion. It invades and ultimately destroys a key white blood cell type that is central to the body's immune defenses, and it is this damage that produces the multiple symptoms of AIDS.

However, death, or even symptoms of

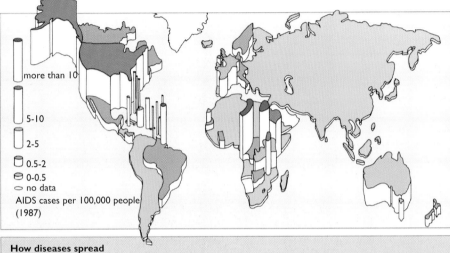

more than 10

5-10

2-5

0.5-2

0-0.5

no data

AIDS cases per 100,000 people (1987)

Number of reported cases in thousands

60

50

40

30

20

10

1987   1988

In 2000, the number of people in North America known to be HIV positive approached the one million mark.

Southern Africa is the area where HIV infection is greatest. Almost 9 percent of the population is infected, including many children who have caught the disease while still in the womb.

How diseases spread

Method of transmission	Transmission agents	Examples
Water	Infective agents (germs or eggs) in drinking water	CHOLERA TYPHOID
Food	Infective agents (germs or eggs) in food	SALMONELLA SOME ROUNDWORMS LISTERIA
Vectors	Insects bite and transmit infection	MALARIA RIVER BLINDNESS SLEEPING SICKNESS
Direct invasion	Active invasion by parasites	HOOKWORMS BILHARZIA
Sex	Disease agent spread during sexual intercourse	SYPHILIS GONORRHEA AIDS
Air	Germs in airborne water droplets	INFLUENZAS COLDS

the disease, can be delayed for up to eight years, during which a carrier of the virus can unwittingly infect other people. This means that the rate at which an AIDS epidemic spreads in a community is closely linked to the rate of sexual partner change. In the United States, the first observable epidemic was among male homosexuals because their average rate of partner change was about 10 times higher than among heterosexuals. In some African societies the primary epidemic has been among heterosexuals, probably because partners changed sufficiently often to ensure a rapid spread of the virus.

The fight against AIDS

Present hopes of curtailing or reversing the epidemic rest first on health education and on employing "safe sex" techniques.

The American homosexual community has already significantly altered its patterns of sexual activity in response to such knowledge. The pharmaceutical industry is also searching for new antiviral drugs that can block the replication of the AIDS virus. AZT was an early, imperfect drug of this type.

Many prototype vaccines against AIDS are also being developed, but success in this area is made more difficult by the ability of the AIDS virus to keep altering the constitution of the proteins at its surface, as the flu virus does. These variable viruses are notoriously difficult to treat with vaccines.

The success of treatment with drugs in reducing deaths from AIDS has lead to some complacency in the use of "safe sex" in the control of its spread.

The first cases of AIDS (acquired immune deficiency syndrome) in the United States were diagnosed in 1981.

The graph (*above left*) shows the number of new cases reported to the World Health Organization from 1980 to 1988. The true total is probably much higher since many cases are thought to remain unreported.

The map (*above*) shows numbers of reported AIDS cases per 100,000 people for all those countries that provide data.

Pollution:
acid rain

The rivers of southern Norway used to yield a rich harvest of Atlantic salmon—66,000 lb in 1900. By 1936 the salmon had virtually died out. It now seems clear that the "pollution cocktail" known as acid rain was responsible, not only for the further depletion of life in Norway's lakes and rivers, which meant that by 1982 1,750 Norwegian lakes had lost all their fish, but also for similar effects on lakes and rivers in other parts of Scandinavia, Europe, and North America.

Acidity is measured on a scale called "pH." A fall in one unit, say from pH7 (neutrality) to pH6, means that acidity is 10 times as strong. At about pH6 such fish as salmon, trout, and roach cannot survive. At pH5, pike and perch die. At pH4.5, even eels die and only some freshwater invertebrates survive.

The death of these animals may be either a direct result of the acidity or a side effect of that acidity that leads to certain toxic elements, such as aluminum, becoming more soluble in lake water.

It is possible to trace the history of acidity in lakes by studying microscopic planktonic organisms, known as diatoms, found in vast quantities in lake sediments. Their hard, highly ornamented, silica coats survive in the mud and can be identified as belonging to particular alkali- or acid-loving diatom species. Diatom analysis of sediments in a Scottish lake revealed a constant pH of 5.7 until 1850; then a decline began that accelerated in the 1920s, leading to a present pH of 4.8. These changes correspond with the increase in industry and air pollution.

The leaching of nutrients over long periods of time, as well as conifer afforestation, could also account for increased soil acidity. But it is now accepted that acidity comes principally from precipitation—in snow, rain, hail, or mist. One compelling piece of evidence is the flush in lake acidity during snow melt in Scandinavia, when a sudden drop in pH causes a peak in fish deaths.

Links between the acidity of rain and the development of industry are easily explained. Burning fossil fuels, such as coal and oil, results in the production of several gases that can dissolve in water to form acids. Sulfur dioxide is probably the worst offender. It is emitted principally from coal-fired power stations and heavy industrial plants, and once in the atmosphere it oxidizes to form sulfuric acid. Oxides of nitrogen, from motor vehicle exhausts, account for at least 50 percent of the nitric acid in the atmosphere.

These pollutants may be carried as far as 1,200 miles before they are deposited. So acids from Britain, Germany, and Czechoslovakia may be washed out by rain over Scandinavia; those from industrial areas in the United States are carried north into Canada. Both Scandinavia and Canada have acid soils that are not able to neutralize the acidic material falling from the sky.

A number of problems have been blamed directly or indirectly on acid rain. It is implicated in the disappearance of certain plant species and a consequent decline in dependent fauna; it is thought to be behind the increase in respiratory diseases and it may be implicated in Alzheimer's disease in humans; it also erodes the stone from which many ancient buildings are constructed.

The dying trees
Potentially the most serious effect of all is forest death, which now affects more than 17 million acres in over 20 countries. It is possible, however, that other factors, such as ozone pollution, periodic drought, mineral deficiency, and infection by pathogens may also be involved here.

The cure for acid rain is not simple.

Sources of pollution
- Homes
- Commerce
- Power stations
- Refineries
- Industry
- Road traffic
- Rail traffic
- Others

Sulfur dioxide

Nitrogen oxides

Sulfur and nitrogen oxides

Dry deposition

Much of the smoke from industrial chimney stacks is in the form of small fragments of solid particles—acidic dust.

Acidic dust lands on plants, clogging their pores, and enters the lungs of animals and humans.

Acidic gases are dispersed downwind of the sites where industrial pollutants are released. Thus much of the acidic gas produced in the industrial north of the United States is blown into Canada before falling as rain.

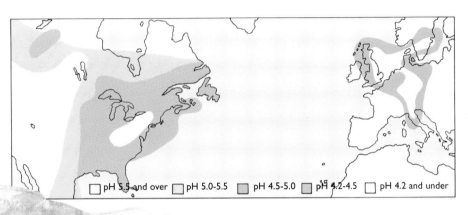

☐ pH 5.5 and over ☐ pH 5.0-5.5 ☐ pH 4.5-5.0 ☐ pH 4.2-4.5 ☐ pH 4.2 and under

As the acidic gases are dispersed by the wind, they are first oxidized by atmospheric oxygen or by ozone, then dissolve in water to form acid droplets in the rain clouds.

Wet deposition—acid rain

Acid particles accumulate in snow; when snow melts it acidifies streams

Direct damage to trees

When acid rain falls on vegetation, it dissolves away the waxy covering of the leaf, leaving cells open to infection and liable to dry out. It may even damage the photosynthetic pigment, chlorophyll.

Lake acidification

Immediate help may be given by adding lime to lakes and raising the pH once again. But the only long-term solution is to cut sulfur emissions, particularly from power stations. Sulfur emissions have been reduced in recent years and the pH of rainfall is beginning to rise. The strongest currently favored contender, nuclear power, also poses environmental problems. Alternative, cleaner forms of energy production include the exploitation of wind, wave, solar, and water power.

Leaching through soil

In the soil the acids dissolve out many plant nutrients and wash them out of the ecosystem. Toxins, such as aluminum, are released from the soil in acid conditions and can damage both the plant life and the fauna of lakes and rivers.

The Parthenon in Athens has suffered more erosion in the past 30 years than over the previous 2,000—clear evidence of the destructive powers of polluted acidic air.

In Germany, the percentage of trees damaged by acid rain rose from 34 to 50 percent between 1983 and 1984, but there has been some recovery since that time as sulfur emissions from fossil fuel burning have been reduced.

Pollution:
the greenhouse effect

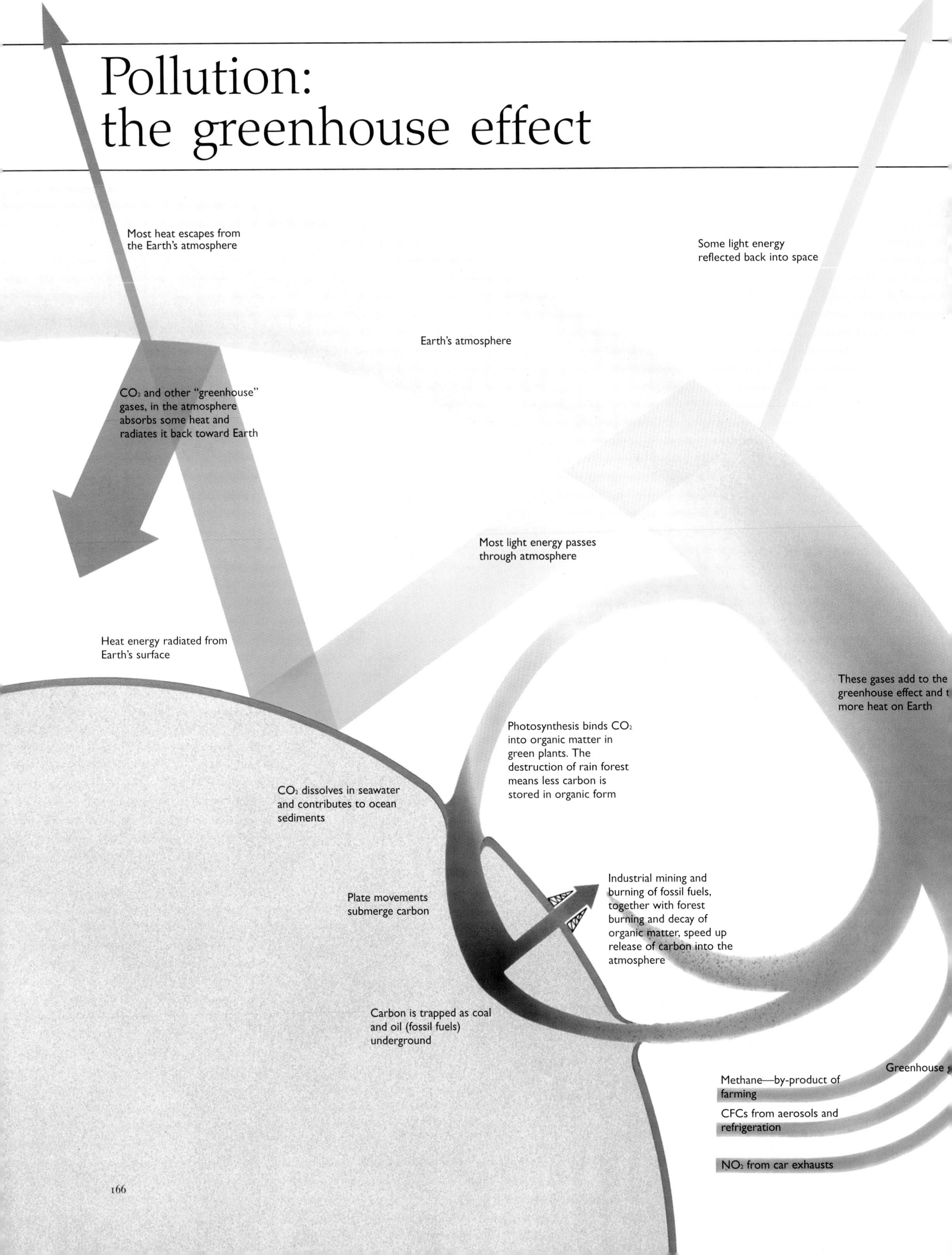

Most heat escapes from
the Earth's atmosphere

Some light energy
reflected back into space

Earth's atmosphere

CO_2 and other "greenhouse"
gases, in the atmosphere
absorbs some heat and
radiates it back toward Earth

Most light energy passes
through atmosphere

Heat energy radiated from
Earth's surface

These gases add to the
greenhouse effect and t
more heat on Earth

Photosynthesis binds CO_2
into organic matter in
green plants. The
destruction of rain forest
means less carbon is
stored in organic form

CO_2 dissolves in seawater
and contributes to ocean
sediments

Industrial mining and
burning of fossil fuels,
together with forest
burning and decay of
organic matter, speed up
release of carbon into the
atmosphere

Plate movements
submerge carbon

Carbon is trapped as coal
and oil (fossil fuels)
underground

Methane—by-product of
farming

Greenhouse

CFCs from aerosols and
refrigeration

NO_2 from car exhausts

e sun

The processes that result in the gradual warming of the Earth's atmosphere, the increasing "greenhouse effect," are illustrated here. The most important of the greenhouse gases is carbon dioxide, and the passage of carbon through the world ecosystem, the "carbon cycle," is also shown.

Solar energy reaches the Earth in the form of short-wavelength light energy. But when that light strikes a surface, much of the energy is converted into long-wavelength heat. Carbon dioxide (CO_2) and other gases present in the atmosphere absorb and retain enough of that radiated heat to maintain a comfortable global temperature. Like the panes of glass in a greenhouse, the gases let the light in, but do not allow heat to escape; hence the "greenhouse effect."

But the concentration of carbon dioxide is increasing and its effect is exacerbated by the presence of rising concentrations of other gases that also absorb infrared radiation efficiently. As more heat is retained the Earth's temperature must rise—perhaps with devastating results. While the world needs its natural "greenhouse," its balance should not be disturbed.

The greenhouse gases

Apart from carbon dioxide, the principal greenhouse gases are water vapor, oxides of nitrogen, ozone, methane, and the chlorofluorocarbons (CFCs). Carbon dioxide is given off in the burning of fossil fuels and in the decay of organic matter in the soils. Methane is produced as a result of bacterial activity associated with certain types of wetland agriculture (such as rice paddies) and the large number of ruminant domestic animals. The CFCs are mainly produced by certain aerosol sprays, refrigerators, and air conditioners and in blowing plastic foams.

Carbon dioxide is responsible for about half this global warming; other pollutants and the other greenhouse gases account for the rest. In 1850, the CO_2 in the atmosphere amounted to 265 parts per million. By 1998 it was 365 ppm. Although still a relatively small quantity of gas, its effect on the average global temperature has been noticeable.

This has risen by almost 2°F over the last century, though the effect has not been felt equally in all parts of the world. Predictions for the future vary considerably, but conservative estimates suggest it may well rise by another 3–4°F higher by the end of this century.

The effect of this on the world's agriculture would be profound, with many more areas becoming desert. And as the ice caps melt, sea levels, according to some scientists, could rise by more than 1 ft by the year 2050.

Increased industrial and domestic combustion of fossil fuels—as well as the burning of large areas of rain forest—have accounted for this rise in CO_2 levels. Some of the excess can be used by plants in photosynthesis and some, being soluble in seawater, can be absorbed by the oceans. But the rate at which it is used up or absorbed cannot keep pace with the rapid rate of CO_2 generation.

Ozone can also act as an atmospheric pollutant and greenhouse gas. This gas is produced at ground level by the interaction of sunlight and the exhaust fumes from motor vehicles, typically producing a "photochemical smog." Ozone is highly reactive and may cause respiratory and eye irritations. It also causes the breakdown of the pigment chlorophyll, thus seriously damaging plant life.

A natural layer of ozone, between about 12 miles and 30 miles high in the atmosphere, shields the Earth from harmful ultraviolet radiation. But this ozone shield is being damaged by the very CFCs that are increasing the greenhouse effect. Since the beginning of the 1980s, it has become particularly thin each spring over Antarctica, forming what is known as an "ozone hole" (see pp. 174–75).

Controlling pollutants

These atmospheric problems can only be tackled by controlling the release of the offending gases. The elimination of certain aerosols, for example, has gone some way to controlling CFC emissions. The increasing demand for refrigerators in tropical countries will tend to offset this, however. Methane is more problematic, since it is still not entirely clear where the major sources are. CO_2 emissions can only be reduced by the control of fossil fuel burning and the conservation and extension of the world's remaining forest cover.

Concern over the thinning of the ozone layer in the Arctic continues to grow.

NASA studies have also pointed to a winter thinning of the ozone layer over some parts of the northern hemisphere of about 8 percent in the last twenty years.

As yet ozone depletion over the Arctic is less severe than that over Antarctica, partly because of the Arctic's higher temperatures. But because the Arctic is so much nearer heavily populated areas, loss of ozone here is a much greater threat to the health of human beings.

As atmospheric CFCs decline it is hoped that the ozone holes will begin to close.

Recent extinctions

The dodo, which became extinct in 1681, can stand as a symbol for much that has happened on Earth since humans became a potent force. It lived on the island of Mauritius, which in the fifteenth century had become a convenient stopping place for spice traders, traveling to and from the East Indies to replenish their stores. The large flightless birds were slaughtered in their thousands for food until, only 200 years after their discovery, there were none left.

The dodo story also serves as a warning that no species is isolated from its environment. Biologists have realized that all the *Calvaria* trees on Mauritius are more than 300 years old; none has taken root since the last dodo died. Since the remains of *Calvaria* nuts are found with dodo skeletons, it seems likely that passage through the digestive system of the dodo was necessary before the nuts could germinate. When eaten by turkeys in experiments they germinate well.

The Maoris of New Zealand also exterminated many bird species, and the Native American are thought to have caused the extinction of two types of Ice Age elephant. It was in the nineteenth century, however, that the problem increased dramatically.

The European expansion through North America in the 1860s was accompanied by an unparalleled slaughter of its wildlife. At that time, for example, the passenger pigeon may have been the most abundant bird in the world; huge flocks could take three days to pass overhead, sometimes at the rate of three million an hour. Less than 100 years later, in 1914, the last passenger died in Cincinnati Zoo.

The slaughter of the bison
The story of the North American Plains bison is equally chilling. There were originally between 30 and 40 million of these creatures, and although the Native Americans killed what they needed for food and hides, the population remained stable. But when European settlers, with their new weapons, swept westward in the 1860s, the bison soon declined in num-

bers. About 2.5 million were killed each year from 1870 to 1875, mainly just for their tongues and hides. Only 500 survived in 1900; since then numbers have increased to more than 25,000 in protected areas.

Human food requirements can no longer be sustained by wild animals, but their slaughter continues for a variety of other reasons. Many are killed for their pelts. The Caspian and Bali tigers are already extinct, and the Australian koala is now rarely seen outside a reserve. The greatest living land mammals are poached for their tusks and horns. Alligators, crocodiles, and snakes are slaughtered for their skins.

Even the sea provides no shelter. In the hunt for the great whales, which provide oil, humans have driven all the larger species to a point where they are endangered or vulnerable to extinction. Hunting of the great Antarctic blue whale began in 1904. By 1963 its numbers had been reduced by 99 percent to 2,000 individuals.

The loss of habitats
Significant alterations to the landscapes of the world have had far-reaching effects. This is not a modern phenomenon, but both the scale and rate of change have increased enormously with the advent of modern technology.

The greatest change is undoubtedly in South America, where the huge Amazon rain forest is being systematically cut down to provide new areas for grazing or cultivation. Since most of the soil is too thin to support crops for more than a few years following such cultivation, erosion sets in and the land crumbles into dust. The Amazon Basin covers 1,200 billion acres and is the most species-rich area in the world. If this devastation goes unchecked, a major storehouse of life's diversity on Earth will disappear forever.

An international document on conservation states that "we have not inherited the Earth from our parents, we have borrowed it from our children." When the loan is returned, it seems likely that it will have been irreparably damaged.

As many as 60,000 plant species may be extinct or near extinct by the middle of the 21st century if present trends continue.

Scientists estimate that for every plant species lost, 10 to 30 other organisms that depended on that plant in some way may also become extinct.

If tropical rain forests continue to be destroyed at the current rate, it is estimated that at least several hundred vertebrates and more than a million species of insect will become extinct within the next 30 to 50 years.

e **African elephant** could be extinct in less than 20 years if poaching continues.

Saving one species: the condor

Big animals, the so-called megafauna, seem always to have been particularly at risk of extinction by human beings. There is strong circumstantial evidence that even in prehistoric times, early human cultures pushed a number of large animals into oblivion, and much current conservation effort is aimed at saving some of those that are now threatened.

These animals are susceptible for many reasons. Their size makes them vulnerable to hunters, since they are easily located and hit with missiles. They need extensive territories, hence populations are rarely large. Litter or clutch size is small, sexual maturation late, and breeding rates often low, so replacement is slow.

In addition, large predators are often sensitive to disturbance, particularly when they are at the top of a food chain, since any changes lower down, such as the loss of a food resource, can prove damaging. And any toxins or pesticides in animals lower down will accumulate in the top predator and may cause death.

Among the most severely endangered large animals is the massive Californian condor. Its dimensions—about 4 ft from head to tail and with a wingspan of 9 ft—place it among the largest flying birds in the world. The condor belongs to the New World vulture family and, like most vultures, is a scavenger of carrion.

A native of the semi-arid scrublands and mountains of southern California, the condor has been in decline for many years. It has been poisoned and persecuted by humans, and increasing efficiency in pastoral agriculture means fewer cattle and sheep carcasses as a source of food.

In the past, the condor's range was much greater: Fossil remains have been found in Texas, New Mexico, and Arizona, where, in the Grand Canyon, there is a clue as to why its distribution became so restricted. Bones of the condor have been found in caves alongside those of many large extinct animals dating from about 10,000–14,000 years ago, and it is evident that these ancient condors scavenged on the dead bodies of the megafauna. With the loss of this source of food, the species began its long decline. In California it may have survived by feeding on the beached bodies of seals and whales.

The condor has all the classic problems of large creatures. It is shy, it lays only one egg, it breeds only on reaching an age of six years, and its food resources are scarce so it needs to range widely. Finally, its association with dead stock animals and its unattractive appearance have given it a poor image and have led to persecution.

When an animal becomes very scarce, it suffers from inbreeding, for its few possible mates may be closely related, and harmful genetic characteristics can accumulate. By 1940 the number of condors had fallen below 100; in 1982 there were only about 20 wild birds left.

The dilemma of intervention

At this point some difficult questions had to be faced. Should the remaining wild birds be fed to encourage population growth? Should their eggs, or even mature birds, be captured to ensure the survival of the species? In 1982 eggs were removed from nests in the wild and incubated in captivity. The young birds were fed with condor-headed puppets to avoid confusing them. As the wild population failed to recover and was clearly beyond rescue, all 27 remaining adult birds were also brought into captivity in 1987. Here they bred very effectively and by 1998 the total captive population exceeded 150 birds.

Release back into the wild commenced in 1992 and by 1999 a total of 88 birds had been released in southern California. A major problem for the released birds is that much of the food they find consists of of animals killed by hunters, and these often contain toxic lead, which can cause poisoning. Hunting must be controlled or the use of lead ammunition banned in areas where condors are being re-established. The annual mortality rate needs to be kept below 10 percent if the wild population is to become viable again.

The Californian condor (*Gymnogyps californianus*), one of the largest flying birds, has approached the edge of extinction.

Nature reserves: selection and design

As human populations increase, even greater pressure than at present will be placed upon the world's decreasing areas of natural habitat. At the same time, it is of great importance that the extinction of species should be avoided; for, apart from all the other reasons for conservation, such a loss of genetic resources is an unacceptable cost to humankind.

If endangered creatures such as the condor and panda are to survive in the wild, nature reserves must be set aside for them. The same applies to smaller and less conspicuous species: habitat maintenance is the key to survival.

In many of the developed countries, unspoiled habitats are already rare, and conservationists are concerned to retain what little survives. In areas that are currently developing, such as the Amazon Basin, there is still an opportunity to set aside sites for nature conservation. But what criteria should be applied to their selection?

Choosing a site

The first priority must be to protect fragile habitats that are under threat. Some of these, such as the raised bogs of Ireland and the swamp forests of Sarawak, have taken many thousands of years to develop, and once lost can never be recreated.

Several other factors must also be taken into account in the selection of a site for special conservation. The diversity of species it contains is important, not only because it permits more species to be cared for in a single reserve, but also because it is sometimes indicative of lack of disturbance in the past. The degree to which an area can be regarded as free from human influence is also significant.

When establishing a reserve within a changing landscape, for example in the Amazon rain forest, the questions of its size and shape and the proximity of other protected areas must be taken into consideration.

In general, the larger the area, the more species it will contain. Moreover, most species that are lost become extinct because of the stresses around the edge of a reserve, and a large site has less perimeter per unit area than a small one. A circular

A **grizzly bear** (*Ursus arctos*) roams the virtually undisturbed territory of a wildlife refuge in Alaska.

shape reduces that perimeter to a minimum. If a reserve is cut back to one-tenth of its original area, probably about half the species it contains will be lost.

In an experimental study in Brazil, forest "islands" were left when an area was cleared and the number of forest-breeding birds recorded a year later. A 3,460-acre site had lost 14 percent of its original birds; a 618-acre site 41 percent; and a 52-acre site as much as 62 percent of its breeding species.

But species number is also affected by diversity of habitat. If there are many microhabitats in an area, it may contain more species than might be expected on the basis of size alone. Some species have specific habitat requirements; for instance, for a few days each year, many species of frogs in the Amazon Basin need wet hollows, preferably those created by wallowing peccaries, in order to breed successfully. Such particular needs must be borne in mind when a reserve is selected.

Multiple reserves

It is also possible to argue that a single large reserve may not be the best answer; in a catastrophe species could be lost. Several smaller reserves, close together, might provide insurance against such loss and allow the recolonization of a devastated area by immigration. Proximity also enhances interbreeding between populations and improves their genetic variation.

The movement of animals from one reserve to another can be encouraged by providing corridors between them. For instance, hedges form corridors for animals between woodland reserves, and in the tropical forest reserves of Costa Rica, corridors have been constructed specifically to permit the seasonal movement of migrant animals between upland and lowland regions.

Although it is possible to formulate guidelines for the selection of ideal nature reserves, the choice of site is often limited. The conservationist may be forced to accept what is available only after other demands on the land have been satisfied.

Nature reserves

The following alphabetical list presents a small sample selection of sites worldwide.

Arctic National Wildlife Refuge ALASKA Virtually undisturbed arctic reserve containing species such as polar and grizzly bears, caribou, musk oxen, golden eagles.

Baikalsky and Darguzinsky State Reserves Russia Include Lake Baikal and contain more than 220 bird species and 40 mammals, such as the Baikal seal, sable, and reindeer.

Dumoga-Bone National Park INDONESIA Contains mammals such as the babirusa, tarsier, Sulawesi civet, and phalanger, 90 percent of which occur nowhere else; outstanding bird fauna, 40 percent of which are found only there.

Eclipse Sound/Bylot Island CANADA Seabird breeding site; also large populations of bowhead and beluga whales, narwhals, seals, walruses, and polar bears.

Fjordland National Park NEW ZEALAND Rare birds such as takahe and kakapo.

Galapagos National Park ECUADOR Birds, such as Darwin's finches, and giant tortoises, marine iguanas.

Gray Whale Lagoons of Baja California MEXICO Major breeding site of gray whales.

Henry Pittier National Park VENEZUELA Huge range of species, particularly birds and insects. Site of spectacular spring and fall migrations of birds and butterflies.

Jau National Park BRAZIL Rain forest reserve containing jaguars, giant otters, primates, macaws, and many other species.

Kanha Tiger Reserve and National Park INDIA Tiger sanctuary also contains sloth bear, wild dog, swamp deer, and others.

Serengeti National Park TANZANIA World's largest community of grassland mammals such as wildebeest, zebra, Thomson's gazelle. Also elephants, black rhinos, lions, cheetahs.

Shark Bay AUSTRALIA Dugongs, turtles, rays, sharks, and shellfish. Islands in bay support many rare mammals such as hare wallabies.

St. Kilda National Nature Reserve UNITED KINGDOM One of the greatest Atlantic seabird colonies. Includes gannets, puffins, fulmars, skuas, and many others.

Swedish Lapland SWEDEN Largest European reserve, includes lynx, bears, wolves, otters, moose, reindeer.

Virunga National Park ZAIRE Mountain gorilla population and perhaps the greatest diversity of species in Africa.

Wenchun Wolong Nature Reserve CHINA Giant pandas and other endangered species such as golden monkey.

Yellowstone National Park USA Grizzly bear, wolf, mountain lion, bald eagle, trumpeter swan, and many others.

Yosemite National Park USA Contains spectacular granite mountains and groves of giant sequoias—among the world's oldest trees; fauna include the black bear, bobcat, and gray fox.

The Great Barrier Reef is the world's most spectacular underwater reserve. It stretches more than 1,250 miles along the northeastern coast of Australia and covers an area of 114,000 sq miles.

The reef contains more than 400 species of coral in a wide variety of forms. They provide a rich habitat for an unequaled collection of marine animals, including more than 1,400 species of fish and 4,000 species of mollusk, as well as myriad sponges, starfish, sea urchins, and many others.

Patterns of the future:
the next 50 years

Predicting what the world will be like in 50 years is a task no scientist would take lightly. Too many unknown factors may play a part in the shaping of the future to forecast it with any accuracy. The current rise in atmospheric temperature caused by the increasing greenhouse effect (see pp. 166–67) could be taken as one example. Predictions can only be made about this by extrapolating from past evidence and making certain assumptions about the continuing accumulation of greenhouse gases in the Earth's atmosphere.

It is possible, though probably unlikely, that our species will modify its behavior and lifestyle sufficiently to slow down the production of carbon dioxide and CFCs. But, equally, it is only too probable that these and other gases will continue to build up even faster than expected.

Will the ozone hole over the Antarctic continue to grow? Will an Arctic equivalent develop and will the frequency of human skin cancers increase accordingly? Or is the ozone hole a seasonal feature of the Antarctic atmosphere that has always existed and has only recently been observed? Will it prove to be of little longterm consequence?

Such questions cannot be answered with certainty. Today, scientists can only make informed guesses about the future and may well be wrong. If, in 2050, some 50 years after the first publication of this book, the authors were asked to write a new edition, what kind of information might it contain? Some possible extracts from such a book follow here.

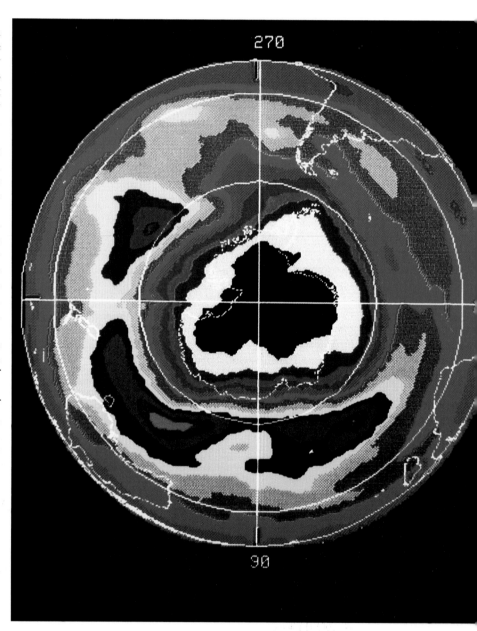

•

Famine in central Africa is still spreading southward as the great drought extends and intensifies. For many years the monsoon rains have failed to penetrate the Sudan, with the result that pastoral agriculture has become entirely unreliable. In West Africa, Nigeria supports refugees from the Sahel who must constantly move farther south with the retreating vegetation.

•

The HLS IV virus is being dubbed the New Plague by the media. From its origin in the suburbs of Mexico City, it has now spread to almost all major centers of population in the world. It is believed to have arisen from a form of the influenza virus, but the high mortality rate associated with infection reflects the present inadequacy of the human immune system to cope with this new mutant strain.

•

Cereal crops are no longer sprayed with chemical pesticides. The transfer of genetic material with specific pest resistance is now routine and the use of hazardous chemicals in agriculture is being phased out. This has encouraged the use of genetically engineered biological control agents as a backup to the newly developed resistant crops. Further development of pesticides has been banned.

•

Civilization goes in circles, it is said, and South America seems to be proving the point. The return to mixed cropping of maize, quinoa, and other crops from the time of the Incas has raised living and nutritional standards for the farmers of the Andes. The World Bank has recently set up a fund to encourage the development of early agricultural practices.

•

The days of the European "grain mountains" are over. All excess production, resulting from the variance of computer predictions at sowing time or unexpectedly light infection of crops by pests, is now automatically cast into the fermenters and used for alcohol production. The use of this environment-friendly fuel in automobiles has considerably reduced the problems of ozone and smog generation that were once so common in great cities such as Los Angeles, New York, and London.

•

Increased efficiency of crop production and protection has greatly reduced the pressure to use all suitable land for agricultural purposes. The area of the United States and Europe given over to nature conservation has doubled since the end of the last century, despite the drought and deteriorating climate in the Wheat Belt Desert of North America. Industrial concerns, such as drug companies and germbank agencies, are now shouldering the bulk of the financial load involved in the maintenance and management of these reserves.

•

Israel has now solved its energy supply problems by piping seawater from the Mediterranean to the Dead Sea, generating hydroelectricity during its descent. The Dead Sea is now also acting as an energy source: the deep saline waters, trapped by less dense waters arriving from the Mediterranean, are heated to temperatures in excess of 140°F by solar radiation and used to drive turbines.

Sea level has risen by 5.4 ft over the past 50 years as a result of the increasing global temperature and the consequent melting of large bodies of ice. The Netherlands has had to abandon extensive areas of previously reclaimed land because of the expense of maintaining sea defenses and the constant pumping of ground water. In Bangladesh, much of the former productive area of the Ganges Delta has also been abandoned to the sea, following a series of devastating floods.

•

It has long been known that some bacteria naturally fix nitrogen in the soil. Supplying this nitrogen to plant roots has recently become much more efficient as a result of inoculating the soil with single-celled animals (protozoans), which feed on the bacteria. The protozoans then deposit the nitrogen near the plant roots, where they are most active, in an easily available form which the plants can utilize freely. Agricultural advisers are recommending the routine inoculation of soils with the appropriate protozoans.

•

Human populations in Africa have stabilized, and the farming of game animals, such as eland, ostrich, Cape buffalo, oryx, and impala, has solved the problem of providing meat; the supply of protein in the continent now seems assured. Farmers also report that many of the former problems caused by tsetse fly and other pests have become much less serious. The native species have an in built resistance to such pests.

•

A new international archaeological group has been formed. It will be chiefly concerned with the recovery and reconstruction of low-lying structures from the last century that have been submerged by –rising sea levels. The main branches of the society will be situated in New York, London, and Tokyo—all cities that have suffered partial inundation as a result of climate change.

525
500
475
450
425
400
375
350
325
300
275
250
225
200
175
150
125

DOBSON UNITS

A satellite picture plotting ozone values reveals an ozone "hole"—a dramatically thinned ozone layer—over the Antarctic area. The ozone hole waxes and wanes but on this particular day, 5 October 1987, it covered 2.5 million sq miles.

In 50 years' time, if conservation measures succeed, it may be possible to record the following: after the reintroduction of Californian condors into the wild, 23 pairs have attempted to breed. A total of 14 young have been fledged and prospects for the future are more hopeful.

For the third year in succession, birds bred in captivity have been released in the Grand Canyon. Conservationists expect, however, that it will be some time before breeding takes place in this region, where it has not been recorded for 10,000 years.

There is still controversy about the provision of carcasses for feeding these released birds; proponents of this "artificial" feeding are being dubbed "wildlife gardeners."

Patterns of the future: the far future

Whatever may happen to our planet over the next 50 million years, it is likely that humans will be here to experience it. We are no less genetically adaptable than other animals, and the extent to which we are able to control our own environment makes it unlikely that we shall be severely threatened, whatever climatic, geographical, or geological changes may take place.

It is also likely that we shall still be recognizably human, for there is no reason to expect any major change in our appearance. Our already vestigial, but sometimes troublesome, appendix will have disappeared, as will our rear molars, or "wisdom" teeth, which already only erupt late in adolescence and often cause trouble. We shall still be erect, intelligent bipeds, and our brains are unlikely to become significantly larger, for there is no correlation between brain size and intelligence.

If our species is at all different, it is more likely to be in its social structure than in its appearance. The greatest social

problems arise from populations that reproduce too rapidly for local resources to support them. In the short term, either famine or war can correct such an imbalance. But in the longer term, changes in the birthrate, produced either by social change or by genetic engineering, will provide the only reliable answer. We shall live longer but have fewer children, so there will be fewer of us worldwide. And the reduction in population densities will mean that our demands on the natural resources of the planet will also be diminished.

The moving continents

But what of the planet itself? The slow but relentless process of plate tectonics will have changed the map of the world. As the Americas continue to move westward, the Atlantic Ocean will widen at the expense of the Pacific. A new, elongate island will lie off the west coast of northern North America, where the movement of the

western Pacific has refted the western part of California northward.

Africa, too, will have moved farther north into Europe, so the Mediterranean Sea will have dried up. The collision of the two continents will have raised a chain of mountains from northwest Africa eastward to Turkey. These mountains will separate Europe from the warmth of northern Africa, and Europe itself will have become cooler. The great farmlands of the center and east will be too cool for the crops they bear today. Only Europe's western margin will still be warmed by the continuing, though weaker, Gulf Stream.

The greatest change will be in the Pacific. Australia will have moved northward, away from its present position in the subtropical region of low rainfall. Instead, it will lie across the equator, where generous rains will enter it from the east. The then much-eroded Great Dividing Range will no longer attract all the rain, and the whole eastern half of

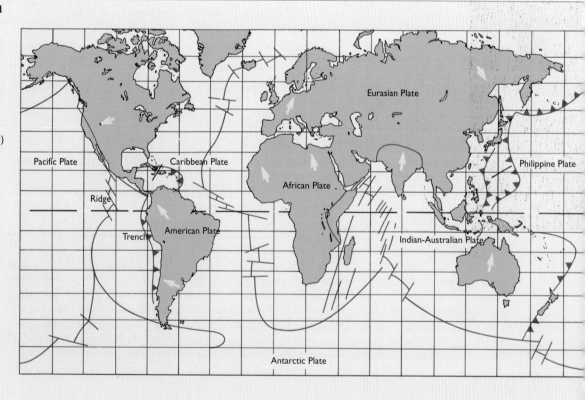

The changing world

The pattern of the Earth's landmasses is not static, and the process of plate tectonics is slowly but steadily changing the shape of the continents.

The first map (*right*) shows the world and its plates as they are today. The second map (*far right*) shows the map of the future and the continental outlines that, on current evidence, are likely to exist in 50 million years' time. Areas of particular interest are circled.

Australia will be wetter and more fertile, though its still impoverished soils will inevitably limit its productivity.

The Antarctic will remain ice-covered, but whether the high latitudes of the northern hemisphere will be affected by alternating glacial and interglacials will depend, as it does today, on a complex series of factors, such as the precise orbit and inclination of the Earth and the level of dust produced by volcanic activity (see pp. 30–31). That, in turn, will have a dramatic effect on sea levels.

If, for example, there is a major ice age in 50 million years' time, the increase in size of the polar ice caps will lead to a reduction in the amount of water in the sea. The sea will, therefore, have receded from the continental shelf extending from Southeast Asia to Borneo; so Australia may well have a complete land connection with Asia, and the present Australian zoogeographic region, with its unique mammals, may have all but vanished under an influx of mammalian species from Asia.

But which mammals will still survive? There can be no doubt that the great mammals will long have become extinct. The great cat predators, the elephants, and rhinos require more space than man can spare. All the productive regions of the Earth will bear crops and support domestic animals—cattle, sheep, pigs, and, in tropical regions, altered varieties of the antelope and buffalo.

The breeding of new crop plants will allow cultivation in cooler and higher latitudes and at greater altitudes. Huge fish-farming enclosures will fringe the eastern Pacific, where a fertile ecosystem will still be based upon the rich, cold upwelling waters near the coasts of the Americas.

Life will be quiet and orderly—as long as humans have completed their last and most important program of domestication: themselves.

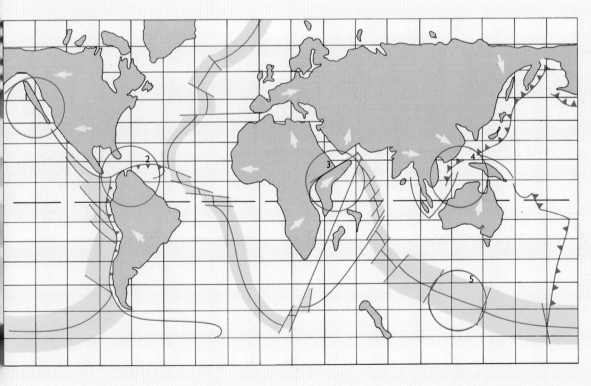

(1) A long silver of the Pacific coast of North America, from San Francisco to Baja California, will move farther north and become an island.

(2) The slow northward movement of South America toward North America will cause the compression of the Central American Panama land bridge.

(3) Today's Great Rift Valley in eastern Africa may extend and eventually split off a fragment of the continent.

(4) (5) The northward movement of the continental plate that underlies Australia and New Guinea will take them into collision with Southeast Asia.

A catalog of life

The Plant Kingdom

The evolution of the Plant Kingdom has resulted in a wide range of different plant types, all linked in the distant past by common ancestors. The accompanying diagram shows the ways in which many botanists believe the different groups of plants are related in an evolutionary series, represented here by branches terminating in the living plant types. The scheme shown is one of a number of possible plant classifications, all of which are based on a consideration of fossil history and modern resemblances between species.

Life began in the oceans and the algae, many of which still inhabit this environment, are regarded as a primitive group of plants, able to survive only when submerged or regularly doused with water. And since they produce swimming sperms, they are totally dependent on water to insure reproduction. Mosses are better equipped to survive in dry environments, but their means of conducting water is primitive and they, too, need water to carry their sperms. Ferns also have swimming sperms, but have developed much

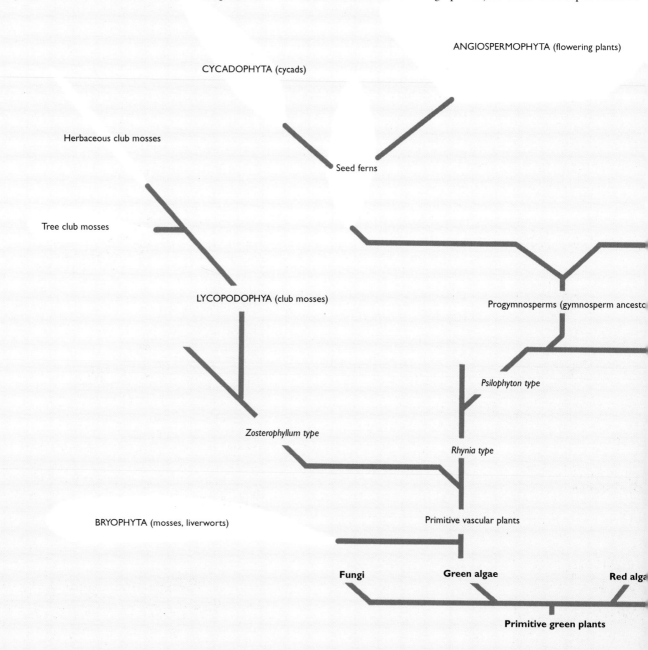

ANGIOSPERMOPHYTA (flowering plants)

CYCADOPHYTA (cycads)

Herbaceous club mosses

Seed ferns

Tree club mosses

LYCOPODOPHYA (club mosses)

Progymnosperms (gymnosperm ancesto

Psilophyton type

Zosterophyllum type

Rhynia type

BRYOPHYTA (mosses, liverworts)

Primitive vascular plants

Fungi   Green algae   Red alga

Primitive green plants

larger and more complex vegetative structures in which water-conducting tissues are present. Some primitive club mosses, close relatives of ferns, were able to attain heights of 130 ft or more by producing such conducting tissue.

Trees first flourished in the conifer group and in the flowering plants (angiosperms). Neither group depends on water to reproduce sexually, and this freedom has undoubtedly played a large part in their success. The earliest flowering plants were probably trees—the herbaceous flowering plants came later in the evolutionary story.

There are still many gaps in the fossil record of plants, such as the jump from algae to mosses and liverworts, and the origin of the flowering plants. It is fairly certain that the green algae are the main ancestral group in both these cases, since the system of chlorophyll pigments is the same throughout the evolutionary series. Other algal groups, such as the brown and red algae, together with the fungi, must have branched off at an earlier stage.

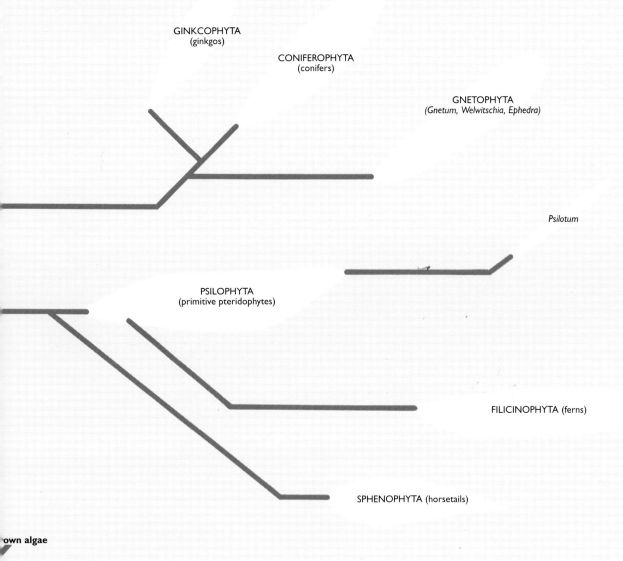

GINKGOPHYTA
(ginkgos)

CONIFEROPHYTA
(conifers)

GNETOPHYTA
(Gnetum, Welwitschia, Ephedra)

Psilotum

PSILOPHYTA
(primitive pteridophytes)

FILICINOPHYTA (ferns)

SPHENOPHYTA (horsetails)

own algae

Flowering plants

A tentative arrangement of evolutionary relationships within the flowering plants, the angiosperms, is illustrated in this diagram. It arranges groups according to their similarities and their presumed evolutionary paths, beginning with an original ancestor whose identity is still unknown. From this ancestor, flowering plants have developed in many different ways, shown here as a radiating series. The main groups are the plant orders, each consisting of several families, but fossil links between orders are scarce or nonexistent. The orders are arranged according to complexity or assumed evolutionary advancement, with the simpler groups toward the center of

the diagram and the more advanced ones at the perimeter.

Some of the oldest flowering plant fossils resemble magnolias in having many floral parts (sepals, petals, stamens, and carpels) arranged spirally. For this reason, plants such as magnolias (Magnoliales) are considered primitive. They have not changed significantly in their floral structure since the early days of the evolution of the flower. The buttercups (Ranunculales), however, are regarded as slightly more advanced. Both their petals and sepals are reduced to a single whorl (sometimes elaborately arranged, as in plants such as delphiniums and monkshood), but they still have a spiral

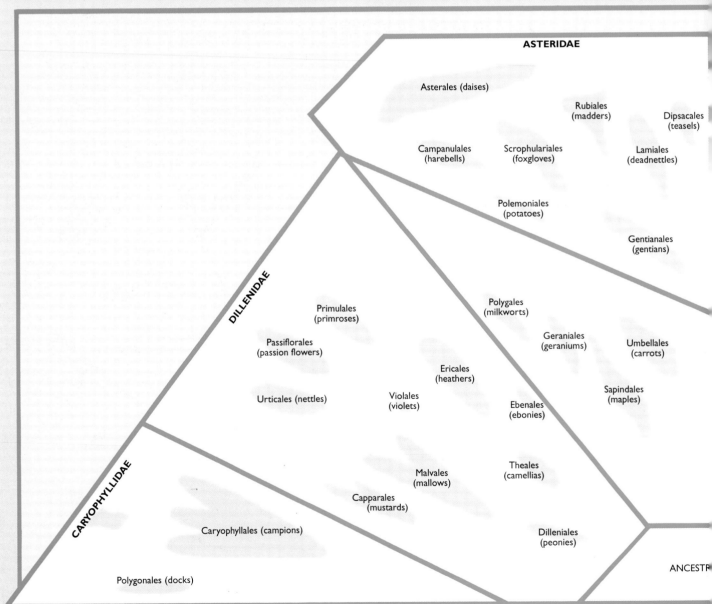

arrangement of many stamens and carpels.

In the more advanced groups of flowering plants, the floral parts are fused into groups. Often petals are joined into tubes—a development related to their pollination by insects. Instead of having a simple, radially symmetrical structure (like that of buttercups), advanced flowers are more complex and often symmetrical about only one plane; snapdragons (Scrophulariales) and sweet peas (Fabales) are examples.

Relatively early on in their evolution the flowering plants developed along two distinct lines, the monocotyledons and the dicotyledons. The two groups differ in four major features: flower form—monocotyledons usually have parts in threes, whereas multiples of five or four are commoner in dicotyledons; leaf venation (dicotyledons have network patterns, while monocots have parallel veins); wood production (absent in almost all monocots); seed germination—dicot seedlings have two seed leaves, whereas monocots have only one.

These two lines of development have diversified greatly, the dicots ranging from oak trees to daisies, the monocots containing such contrasting forms as palms, orchids, daffodils, and grasses.

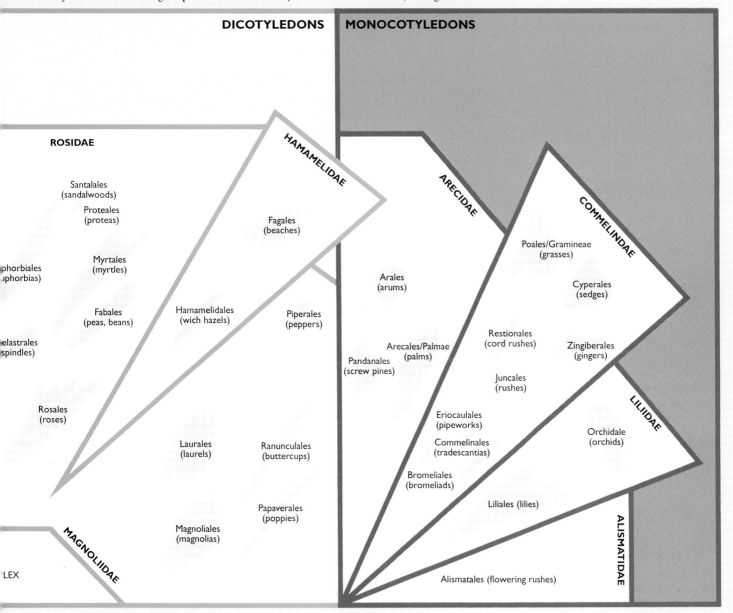

The Animal Kingdom

The basic unit of classification in the Animal Kingdom is the species—a natural grouping of animals that can all breed successfully with each other. Species that share several common characteristics are grouped together into a genus, genera into families, and so on. In ascending order, the most commonly used levels of classification are species, genus, family, order, class, and phylum. Ultimately all animals are grouped together in the Kingdom Animalia.

This evolutionary tree of the Animal Kingdom shows the relationships between all the different groups of living creatures. This diagram links every living animal from the simplest, single-celled animal ancestors, which first lived more than a billion years ago, through to all the major groupings (phyla) of today.

One of the dozens of phyla in the Animal Kingdom contains the vertebrates—animals with backbones, like

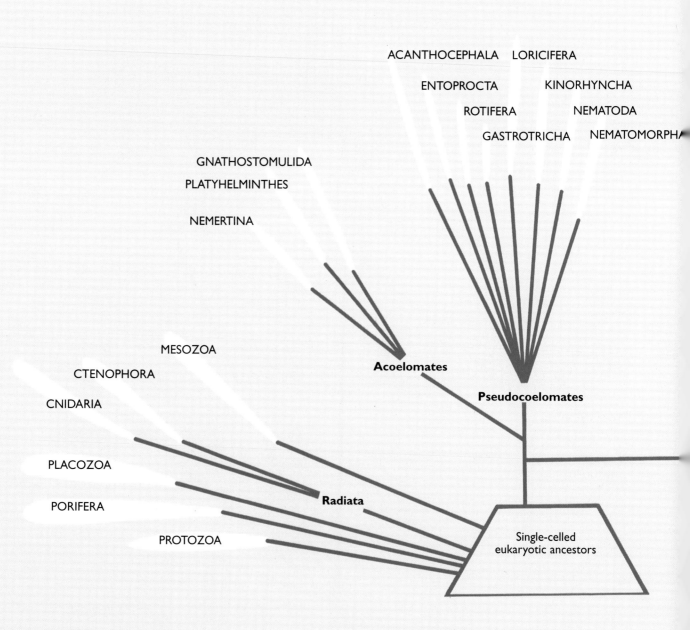

ACANTHOCEPHALA   LORICIFERA

ENTOPROCTA   KINORHYNCHA

ROTIFERA   NEMATODA

GASTROTRICHA   NEMATOMORPH/

GNATHOSTOMULIDA

PLATYHELMINTHES

NEMERTINA

MESOZOA

CTENOPHORA

CNIDARIA

PLACOZOA

PORIFERA

PROTOZOA

Acoelomates

Pseudocoelomates

Radiata

Single-celled
eukaryotic ancestors

ourselves. The vertebrate branch of the tree is examined in greater detail on the next page.

In reading this tree, from its billion-year-old roots to its present-day phylum branch ends, we must bear two points in mind. First, it omits major groups that are now extinct. There are many of these, such as the graptolites which disappeared from the Earth millions of years ago.

Second, any such tree can only be constructed on the basis of "best guesses." The links these suggest all occurred many millions of years ago and no direct evidence for them survives. There is, however, a considerable body of knowledge based on indirect evidence—fossils and the genetic and physical structure of living animals—that is constantly expanding. Minor adjustments to the accepted view of the Animal Kingdom are continually being made in the light of changing knowledge.

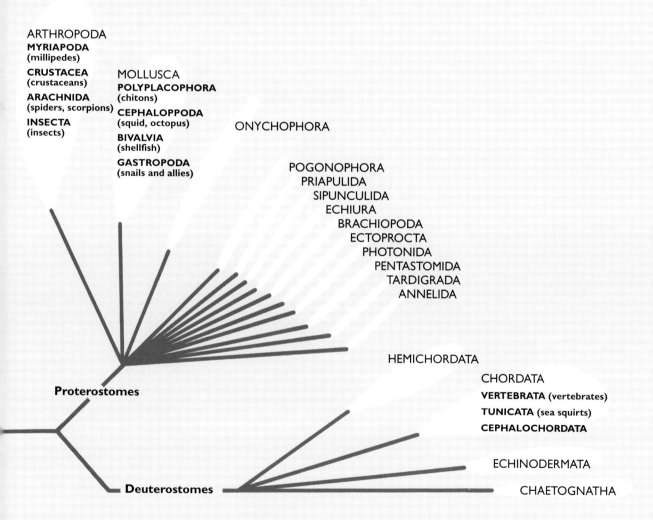

ARTHROPODA
MYRIAPODA
(millipedes)
CRUSTACEA
(crustaceans)
ARACHNIDA
(spiders, scorpions)
INSECTA
(insects)

MOLLUSCA
POLYPLACOPHORA
(chitons)
CEPHALOPPODA
(squid, octopus)
BIVALVIA
(shellfish)
GASTROPODA
(snails and allies)

ONYCHOPHORA

POGONOPHORA
PRIAPULIDA
SIPUNCULIDA
ECHIURA
BRACHIOPODA
ECTOPROCTA
PHOTONIDA
PENTASTOMIDA
TARDIGRADA
ANNELIDA

HEMICHORDATA

CHORDATA
VERTEBRATA (vertebrates)
TUNICATA (sea squirts)
CEPHALOCHORDATA

ECHINODERMATA

CHAETOGNATHA

Proterostomes

Deuterostomes

Vertebrate animals

The vertebrate branch of the chordate phylum in the evolutionary tree of the Animal Kingdom is illustrated here in greater detail than the other phyla. The diagram shows the various classes included in the group—mammals, birds, fishes, and so on—and the orders contained in those classes. Within these orders are families and then species, too numerous to mention here.

Vertebrates have been an extremely successful group. While invertebrate groups may contain more species and exist in greater numbers of individuals, the largest, most complex animals on land, in the air, or in water are almost always vertebrates.

The diagram does not include extinct groupings—there are no dinosaurs. And, despite the fact that the vertebrates are the best known of all animals, there is sometimes room for doubt about origins and links between the different groups. The ways in which the reptiles, mammals, and birds relate to one another, for example, are still a matter of dispute. This tree simply offers one interpretation of modern evidence and knowledge.

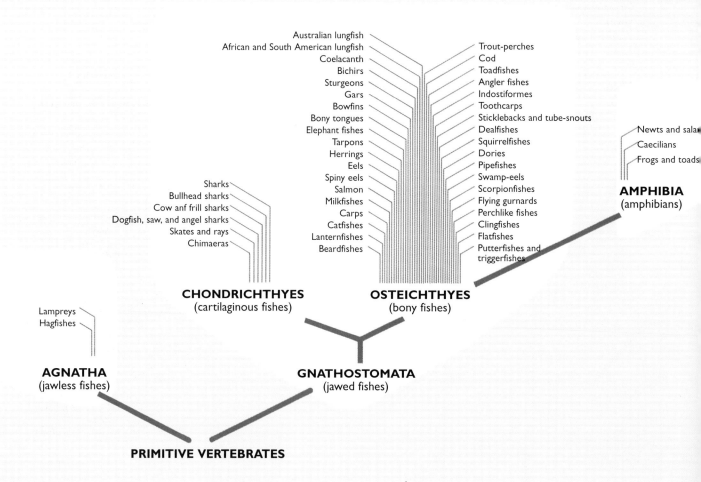

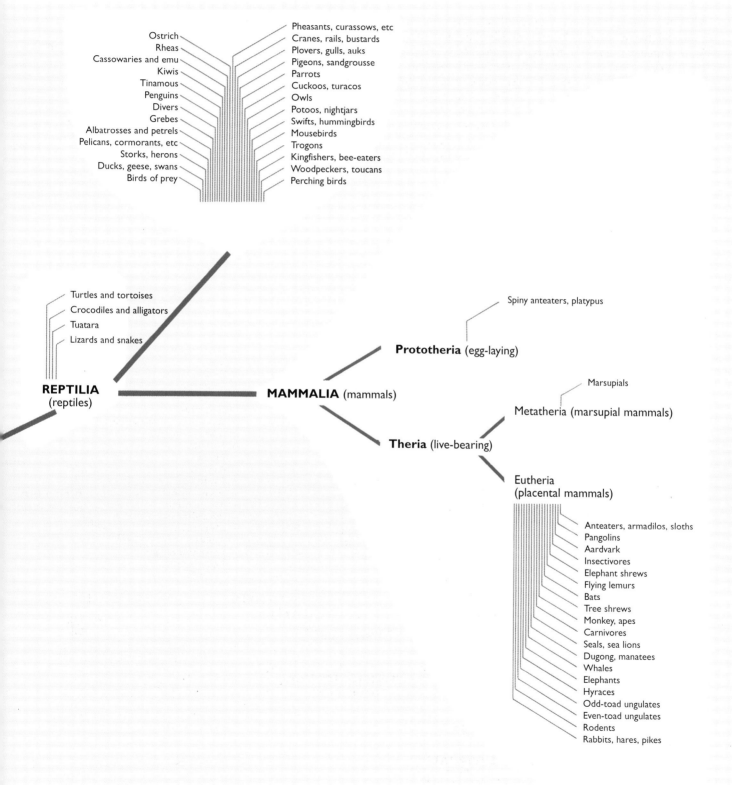

Ostrich
Rheas
Cassowaries and emu
Kiwis
Tinamous
Penguins
Divers
Grebes
Albatrosses and petrels
Pelicans, cormorants, etc
Storks, herons
Ducks, geese, swans
Birds of prey

Pheasants, curassows, etc
Cranes, rails, bustards
Plovers, gulls, auks
Pigeons, sandgrousse
Parrots
Cuckoos, turacos
Owls
Potoos, nightjars
Swifts, hummingbirds
Mousebirds
Trogons
Kingfishers, bee-eaters
Woodpeckers, toucans
Perching birds

Turtles and tortoises
Crocodiles and alligators
Tuatara
Lizards and snakes

REPTILIA
(reptiles)

MAMMALIA (mammals)

Prototheria (egg-laying)

Spiny anteaters, platypus

Theria (live-bearing)

Marsupials

Metatheria (marsupial mammals)

Eutheria
(placental mammals)

Anteaters, armadilos, sloths
Pangolins
Aardvark
Insectivores
Elephant shrews
Flying lemurs
Bats
Tree shrews
Monkey, apes
Carnivores
Seals, sea lions
Dugong, manatees
Whales
Elephants
Hyraces
Odd-toad ungulates
Even-toad ungulates
Rodents
Rabbits, hares, pikes

Successful families: plants

The number of species in families of animals and plants is one indication of their relative success, showing how well that family has adapted to a wide range of specific lifestyles. Generally, the more species there are in a family, the more adaptive potential there is in that basic living design.

The biggest families also tend to have a wide distribution. The distributions of some of the most successful plants, mammals, and birds are shown over the following pages. In each group, families are arranged in descending order according to the number of species they contain.

Sunflowers

Family: Compositae
Species: about 25,000 **Genera:** about 1,100
Examples: lettuce, tarragon, marigold

Orchids

Family: Orchidaceae
Species: about 18,000 **Genera:** about 750
Examples: vanilla, slipper orchid

Peas

Family: Leguminosae
Species: 17,000 **Genera:** about 700
Examples: pea, lentil, chick pea, clover

Grasses

Family: Gramineae
Species: about 9,000 **Genera:** about 650
Examples: wheat (*Triticum* spp.), rice, oats

Gardenias

Family: Rubiaceae
Species: about 7,000 **Genera:** about 500
Examples: gardenia, coffee, quinine

Spurges

Family: Euphorbiaceae
Species: over 5,000 **Genera:** about 300
Examples: rubber tree, poinsettia

Reeds and sedges

Family: Cyperaceae
Species: about 4,000 **Genera:** about 90
Examples: papyrus (*Cyperus papyrus*)

Lilies

Family: Liliaceae
Species: about 3,500 **Genera:** about 250
Examples: tulip, hyacinth, onion

Roses

Family: Rosaceae
Species: about 3,370 **Genera:** about 122
Examples: rose, *Potentilla*, apple, *Prunus*

Mustard

Family: Cruciferae
Species: about 3,000 **Genera:** about 380
Examples: mustard, cabbage, wallflower

Heathers

Family: Ericaceae
Species: about 3,000 **Genera:** about 100
Examples: heathers (*Erica* spp.), cranberry

Mint

Family: Labiatae
Species: about 3,000 **Genera:** about 200
Examples: mint (*Mentha* spp.), marjoram

Disotis and Medinilla

Family: Melastomataceae
Species: about 3,000 **Genera:** about 240
Examples: pococa, maieta

Figs

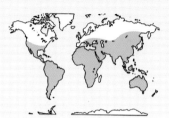

Family: Moraceae
Species: about 3,000 **Genera:** 75
Examples: fig (*Ficus* spp.), mulberry,
breadfruit, iroko

Myrtles

Family: Myrtaceae
Species: about 3,000 **Genera:** about 100
Examples: *Eucalyptus*, clove, pimento, guava,
myrtle

Foxgloves

Family: Scrophulariaceae
Species: about 3,000 **Genera:** about 220
Examples: foxglove (*Digitalis* spp.),
snapdragon, speedwell, *Mimulus, Nemesia*

Verbenas

Family: Verbenaceae
Species: about 3,000 **Genera:** about 75
Examples: lemon verbena (*Lippia citriodora*),
teak, vervain, *Lantana*

Carrot

Family: Umbelliferae
Species: 2,500–3,000 **Genera:** about 300
Examples: carrot (*Daucus carota*), parsnip,
celery, parsley, fennel, anise, hemlock

Potatoes

Family: Solanaceae
Species: 2,000–3,000 **Genera:** about 90
Examples: potato (*Solanum tuberosum*),
eggplant, pepper (*Capsicum*), tomato

Palms

Family: Palmae
Species: about 2,780 **Genera:** about 212
Examples: date palm, coconut, raffia palm, oil
palm, rattan (*Calamus*)

Acanthus

Family: Acanthaceae
Species: about 2,500 **Genera:** about 250
Examples: black-eyed Susan (*Thunbergia*),
Aphelandra, Acanthus

Avocado

Family: Lauraceae
Species: about 2,500 **Genera:** about 32
Examples: avocado (*Persea*), cinnamon,
Sassafras, bay laurel

Mesembryanthemum

Family: Aizoaceae
Species: about, 2,300 **Genera:** about 143
Examples: pebble plants (*Lithops*), ice plant
(*Mesembryanthemum*), *Lampranthus*

Cactus

Family: Cactaceae
Species: over 2,000 **Genera:** 87
Examples: prickly pear (*Opuntia*),
Mammillaria, Melocactus, Ferocactus

Elms

Family: Ulmaceae
Species: over 2,000 **Genera:** 16
Examples: white elm (*Ulmus laevis*), nettle-
tree (*Celtis*), zelkova

Soursops

Family: Annonaceae
Species: about 2,000 **Genera:** about 120
Examples: soursop (*Annona squamosa*),
sweetsop, ilang-ilang

Aroids

Family: Araceae
Species: about 2,000 **Genera:** about 110
Examples: arum lily, *Monstera deliciosa,
Philodendron, Dieffenbachia*

Forget-me-nots

Family: Boraginaceae
Species: about 2,000 **Genera:** about 100
Examples: forget-me-not (*Myosotis*),
heliotrope, comfrey, borage

Pineapple

Family: Bromeliaceae
Species: about 2,000 **Genera:** about 50
Examples: pineapple (*Ananas comosus*),
Spanish moss, *Billbergia, Aechmea, Bromelia*

Carnations

Family: Caryophyllaceae
Species: about 2,000 **Genera:** about 80
Examples: carnation (*Dianthus caryophyllus*),
campion, *Gypsophila*, chickweed

African violets

Family: Gesneriaceae
Species: about 2,000 **Genera:** about 125
Examples: African violet (*Saintpaulia*),
gloxinia, *Achimenes*

Pepper

Family: Piperaceae
Species: about 2,000 **Genera:** about 5
Examples: pepper (*Piper nigrum*), kava

Akees

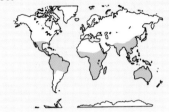

Family: Sapindaceae
Species: about 2,000 **Genera:** about 150
Examples: akee (*Blighia sapida*), lychee,
rambutan, balloon vine

Periwinkles

Family: Apocynaceae
Species: about 1,500 **Genera:** about 180
Examples: periwinkle (*Vinca*), oleander,
frangipani

Milkweeds

Family: Asclepiadaceae
Species: 1,800–2,000 **Genera:** about 250
Examples: milkweed (*Asclepias*), wax plant,
Stephanotis

Buttercups

Family: Ranunculaceae
Species: over 1,800 **Genera:** about 50
Examples: buttercup (*Ranunculus*), celandine,
larkspur, *Pulsatilla*, clematis

Bindweeds

Family: Convolvulaceae
Species: about 1,800 **Genera:** about 50
Examples: bindweed (*Convolvulus*), sweet
potato, morning glory

Iris

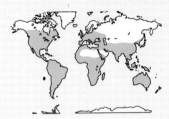

Family: Iridaceae
Species: about 1,800 **Genera:** about 70
Examples: iriis, crocus, gladiolus, freesia,
monbretia

Sugar beet

Family: Chenopodiaceae
Species: about 1,500 **Genera:** about 100
Examples: sugar beet (*Beta vulgaris*), quinoa,
spinach, beet

Stonecrops

Family: Crassulaceae
Species: about 1,500 **Genera:** about 35
Examples: stonecrop (*Sedum*), houseleeks,
Echeveria, Kalanchoe

Mistletoes

Family: Loranthaceae
Species: about 1,300 **Genera:** about 35
Examples: mistletoe (*Viscum/Phoradendron*),
Loranthus

Ginger

Family: Zingiberaceae
Species: about 1,300 **Genera:** about 49
Examples: ginger (*Zingiber officinale*),
cardamom, turmeric, ginger lily

West Indian boxwood

Family: Flacourtiaceae
Species: about 1,250 **Genera:** about 89
Examples: boxwood (*Gossypiospermum praecox*), *Homalium*

Currants and saxifrages

Family: Saxifragaceae
Species: about 1,250 **Genera:** about 80
Examples: black, white, and red currants (*Ribes*), gooseberries, saxifrage, hydrangea

Pipeworts

Family: Eriocaulaceae
Species: about 1,200 **Genera:** 13
Examples: *Eriocaulon, Paepalanthus, Sygonanthus, Leiothrix*

Lobelias

Family: Lobeliaceae
Species: about 1,200 **Genera:** about 30
Examples: lobelia (*Lobelia erinus*), *Clermontia, Cyanea*

Daffodils

Family: Amaryllidaceae
Species: about 1,100 **Genera:** about 75
Examples: daffodils (*Narcissus*), snowdrop, amaryllis, *Nerine, Clivia*

Tea

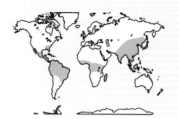

Family: Theaceae
Species: about 1,100 **Genera:** about 29
Examples: tea (*Camellia sinensis*), *Franklinia*, camellia

Cotton and mallows

Family: Malvaceae
Species: over 1,000 **Genera:** about 80
Examples: cotton (*Gossypium*), okra, mallow, hollyhock, *Hibiscus*

Stinging nettles

Family: Urticaceae
Species: over 1,000 **Genera:** about 45
Examples: stinging nettle (*Urtica dioica*), ramie, *Cecropia*

Beeches and oaks

Family: Fagaceae
Species: about 1,000 **Genera:** about 8
Examples: beeches (*Fagus*), oaks (*Quercus*), chestnut (*Castanea*)

Mangosteens

Family: Guttiferae
Species: about 1,000 **Genera:** about 40
Examples: mangosteen (*Garcinia mangostana*), mammee apple, hypericum

Myrsinaceae

Family: Myrsinaceae
Species: about 1,000 **Genera:** about 32
Examples: *Myrsine, Ardisia, Maesa, Suttonia*

Pepper elders

Family: Peperomiaceae
Species: about 1,000 **Genera:** 4
Examples: pepper elder (*Peperomia*), *Manekia, Piperanthera, Verhuellia*

Milkworts

Family: Polygalaceae
Species: about 1,000 **Genera:** about 17
Examples: *Polygala, Xanthophyllum, Bredemeyera, Monnina*

Proteas

Family: Proteaceae
Species: about 1,000 **Genera:** 62
Examples: giant protea (*Protea cynaroides*), *Banksia*, Chilean fire bush, macadamia

Primrose

Family: Primulaceae
Species: almost 1,000 **Genera:** 28
Examples: primroses (*Primula*), cyclamen, auricula

Successful families: mammals

Rats and mice

Family: Muridae **Order:** Rodentia
Number of species: about 1,011
Examples: house rat, harvest mouse, North African gerbil, muskrat, bog lemming

Vespertilionid bats

Family: Vespertilionidae **Order:** Chiroptera
Number of species: about 319
Examples: long-eared myotis, brown pipistrelle, noctule, red bat, barbastelle

Squirrels

Family: Sciuridae **Order:** Rodentia
Number of species: about 246
Examples: gray squirrel, African palm squirrel, eastern chipmunk, woodchuck

Shrews

Family: Soricidae **Order:** Insectivora
Number of species: about 246
Examples: short-tailed shrew, pygmy white-toothed shrew, mouse shrew

Fruit bats

Family: Pteropodidae **Order:** Chiroptera
Number of species: about 173
Examples: greater fruit bat, hammerheaded bat, large flying fox

New World leaf-nosed bats

Family: Phyllostomatidae **Order:** Chiroptera
Number of species: about 140
Examples: little big-eared bat, false vampire, yellow-shouldered bat

Cattle, antelope, sheep, goats

Family: Bovidae **Order:** Artiodactyia
Number of species: about 123
Examples: bison, roan antelope, wildebeest, gazelle, mountain goat, barbary sheep

Free-tailed bats

Family: Molossidae **Order:** Chiroptera
Number of species: about 91
Examples: Egyptian free-tailed bat, black mastiff bat, wrinkle-lipped bat

Old World monkeys

Family: Cercopithecidae **Order:** Primates
Number of species: about 76
Examples: stump-tailed macaque, olive baboon, vervet monkey, red colobus monkey, langur

Opossums

Family: Didelphidae **Order:** Marsupialia
Number of species: about 73
Examples: Virginia opossum, water opossum, mouse-opossum, gray short-tailed opossum

Civets

Family: Viverridae **Order:** Carnivora
Number of species: about 72
Examples: African linsang, masked palm civet, small-spotted genet, meerkat, Indian mongoose

Horseshoe bats

Family: Rhinolophidae **Order:** Chiroptera
Number of species: about 69
Examples: greater horseshoe bat, Philippine horseshoe bat

Weasels

Family: Mustelidae **Order:** Carnivora
Number of species: about 67
Examples: marten, western polecat, least
weasel, wolverine, badger

Pocket mice

Family: Heteromyidae **Order:** Rodentia
Number of species: about 63
Examples: silky pocket mouse, pale kangaroo
mouse, desert kangaroo rat

Old World leaf-nosed bats

Family: Hipposideridae **Order:** Chiroptera
Number of species: about 61
Examples: flower-faced bat, trident bat, large
Malay leaf-nosed bat

Kangaroos

Family: Macropodidae **Order:** Marsupialia
Number of species: about 57
Examples: musk rat-kangaroo, potoroo,
swamp wallaby, red kangaroo

Sheath-tailed bats

Family: Emballonuridae **Order:** Chiroptera
Number of species: about 50
Examples: African sheath-tailed bat, lesser
white-lined bat, tomb bat, ghost bat

American spiny rats

Family: Echimyidae **Order:** Rodentia
Number of species: about 45
Examples: armored rat, gliding spiny rat,
guira, coro-coro

Rabbits, hares

Family: Leporidae **Order:** Lagomorpha
Number of species: about 40
Examples: brown hare, black-tailed
jackrabbit, desert cottontail, common rabbit

Pocket gophers

Family: Geomyidae **Order:** Rodentia
Number of species: about 37
Examples: Plains pocket gopher, northern
pocket gopher

Dogs, foxes

Family: Canidae **Order:** Carnivora
Number of species: about 35
Examples: golden jackal, coyote, gray wolf,
red fox, fennec fox, hunting dog

Cats

Family: Felidae **Order:** Carnivora
Number of species: about 35
Examples: wild cat, lynx, ocelot, cheetah, lion,
jaguar, leopard, tiger

Deer

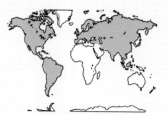

Family: Cervidae **Order:** Artiodactyla
Number of species: about 34
Examples: muntjac, roe deer, tufted deer,
moose, wapiti, caribou, northern pudu

Tenrecs

Family: Tenrecidae **Order:** Insectivora
Number of species: about 33
Examples: greater hedgehog tenrec, streaked
tenrec, shrew tenrec, giant otter shrew

New World monkeys

Family: Cebidae **Order:** Primates
Number of species: about 32
Examples: black-bearded saki, white-fronted
capuchin, red howler, woolly spider monkey

Marine dolphins

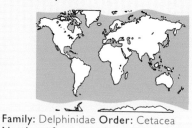

Family: Delphinidae **Order:** Cetacea
Number of species: about 32
Examples: common dolphin, killer whale,
long-finned pilot whale, bottle-nosed dolphin

Tuco-tucos

Family: Ctenomyidae **Order:** Rodentia
Number of species: about 32
Examples: tuco-tuco

Moles

Family: Talpidae **Order:** Insectivora
Number of species: about 29
Examples: European mole, star-nosed mole, Chinese shrew-mole, Russian desman

Jerboas

Family: Dipodidae **Order:** Rodentia
Number of species: about 27
Examples: northern three-toed jerboa, greater fat-tailed jerboa

Ringtalls

Family: Petauridae **Order:** Marsupialia
Number of species: about 22
Examples: sugar glider, common ringtail, common striped possum

Armadillos

Family: Dasypodidae **Order:** Edentata
Number of species: about 20
Examples: nine-banded armadillo, pink fairy armadillo, pichis

Bandicoots

Family: Peramelidae **Order:** Marsupialia
Number of species: about 17
Examples: brown bandicoot, Gunn's bandicoot, mouse-bandicoot

Seals

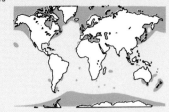

Family: Phocidae **Order:** Pinnipedia
Number of species: about 19
Examples: harp seal, leopard seal, northern elephant seal, Mediterranean monk seal

Raccoons

Family: Procyonidae **Order:** Carnivora
Number of species: about 18
Examples: raccoon, coati, kinkajou, olingo, coatimundi, ringtail/cacomistle

Beaked whales

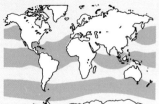

Family: Ziphiidae **Order:** Cetacea
Number of species: about 18
Examples: northern bottle-nosed whale, Cuvier's beaked whale

Golden moles

Family: Chrysochloridae **Order:** Insectivora
Number of species: about 17
Examples: Cape golden mole, hottentot golden mole, giant golden mole

Hedgehogs

Family: Erinaceidae **Order:** Insectivora
Number of species: about 17
Examples: western European hedgehog, desert hedgehog, moonrat, shrew-hedgehog

Marmosets

Family: Callitrichidae **Order:** Primates
Number of species: about 17
Examples: Goeldi's marmoset, golden lion tamarin, emperor tamarin

Tree shrews

Family: Tupaiidae **Order:** Scandentia
Number of species: about 16
Examples: common tree shrew, feather-tailed tree shrew

Lemurs

Family: Lemuridae **Order:** Primates
Number of species: about 16
Examples: ring-tailed lemur, ruffed lemur, gray gentle lemur

Guinea pigs

Family: Caviidae **Order:** Rodentia
Number of species: about 15
Examples: cavy, rock cavy, mara, guinea pig

Elephant-shrews

Family: Macroscelididae **Order:** Macroscelidea
Number of species: about 15
Examples: short-eared elephant shrew

Sea lions

Family: Otariidae **Order:** Pinnipedia
Number of species: about 14
Examples: California sea lion, northern fur seal, northern sea lion

Dormice

Family: Gliridae **Order:** Rodentia
Number of species: about 14
Examples: fat dormouse, Japanese dormouse

Pikas

Family: Ochotonidae **Order:** Lagomorpha
Number of species: about 14
Examples: northern pika, steppe pika, Chinese red pika

Lorises

Family: Lorisidae **Order:** Primates
Number of species: about 12
Examples: slender loris, potto, angwantibo, greater bushbaby, Demidoff's galago

Old World porcupines

Family: Hystricidae **Order:** Rodentia
Number of species: about 12
Examples: Indonesian porcupine, crested porcupine, Asian brush-tailed porcupine

Pacas and agoutis

Family: Dasyproctidae **Order:** Rodentia
Number of species: about 12
Examples: paca, Brazilian agouti, gray agouti, acuchi

Hutias

Family: Capromyidae **Order:** Rodentia
Number of species: about 12
Examples: Ingraham's hutia, coypu/nutria

Phalangers

Family: Phalangeridae **Order:** Marsupialia
Number of species: about 11
Examples: brush-tailed possum, common phalanger, silky phalanger

Slit-faced bats

Family: Nycteridae **Order:** Chiroptera
Number of species: about 11
Examples: Egyptian slit-faced bat, large slit-faced bat

Apes

Family: Pongidae **Order:** Primates
Number of species: about 10
Examples: hoolock gibbon, siamang, orangutan, chimpanzee, gorilla

Jumping mice

Family: Zapodidae **Order:** Rodentia
Number of species: about 10
Examples: meadow jumping mouse, northern birch mouse

New World porcupines

Family: Erithizontidae **Order:** Rodentia
Number of species: about 9
Examples: North American porcupine

African mole-rats

Family: Bathyergidae **Order:** Rodentia
Number of species: about 9
Examples: Cape mole-rat, naked mole-rat, silvery mole-rat

Horses

Family: Equidae **Order:** Perissodactyla
Number of species: about 8
Examples: Przewaiski's wild horse, common zebra, African ass, kiang, onager

Pigs

Family: Suidae **Order:** Artiodactyla
Number of species: about 8
Examples: bush pig, warthog, wild boar, giant forest hog, babirusa

Successful families: birds

Thrushes and allies

Family: Muscicapidae **Order:** Passeriformes
Number of species: about 1,394
Examples: robin, nightingale, wheatear,
spotted flycatcher, willow warbler

Buntings

Family: Emberizidae **Order:** Passeriformes
Number of species: about 569
Examples: yellowhammer, dark-eyed junco,
rufous-sided towhee, cardinal

Tyrant flycatchers

Family: Tyrannidae **Order:** Passeriformes
Number of species: about 370
Examples: eastern kingbird, kiskadee
flycatcher, eastern wood peewee

Hummingbirds

Family: Trochilidae **Order:** Apodiformes
Number of species: about 319
Examples: bee hummingbird, frilled
coquette, marvelous spatule-tail, Andean
hillstar

Pigeons

Family: Columbidae **Order:** Columbiformes
Number of species: about 295
Examples: ground dove, bar-tailed pigeon,
collared dove, superb fruit dove

Parrots

Family: Psittacidae **Order:** Psittaciformes
Number of species: about 243
Examples: kea, gray parrot, peach-faced
lovebird, kakapo, scarlet macaw, budgerigar,
crimson rosella

Antbirds

Family: Formicariidae **Order:** Passeriformes
Number of species: about 230
Examples: ocellated antbird, streaked
antwren, great antshrike, chestnut-crowned
antpitta

Ovenbirds

Family: Furnariidae **Order:** Passeriformes
Number of species: about 219
Examples: plain xenops, rufous
ovenbird, red-faced spinetail, larklike
bushrunner

Hawks

Family: Accipitridae **Order:** Falconiformes
Number of species: about 217
Examples: lappet-faced vulture, red kite,
bald eagle, goshawk, buzzard, golden eagle

Pheasants

Family: Phasianidae **Order:** Galliformes
Number of species: about 214
Examples: European quail, golden pheasant,
Congo peacock, black grouse, turkey

Woodpeckers

Family: Picidae **Order:** Piciformes
Number of species: about 208
Examples: wryneck, great spotted
woodpecker, yellow-bellied sapsucker,
common flicker

Honeyeaters

Family: Meliphagidae **Order:** Passeriformes
Number of species: about 167
Examples: brown honeyeater, Kauai o-o,
little friarbird, tui, Cape sugarbird, red
wattlebird

Finches

Family: Fringillidae **Order:** Passeriformes
Number of species: about 153
Examples: chaffinch, canary, red crossbill, pine grosbeak, common redpoll, bullfinch

Weavers

Family: Ploceidae **Order:** Passeriformes
Number of species: about 143
Examples: village weaver, red-billed quelea, house sparrow, snow finch, paradise whydah

Ducks

Family: Anatidae **Order:** Anseriformes
Number of species: about 140
Examples: Tundra swan, Canada goose, shelduck, mallard, red-breasted merganser

Owls

Family: Strigidae **Order:** Strigiformes
Number of species: about 134
Examples: snowy owl, long-eared owl, tawny owl, morepork, hawk owl

Rails

Family: Rallidae **Order:** Gruiformes
Number of species: about 132
Examples: takahe, common moorhen, water rail, American coot, corncrake

Cuckoos

Family: Cuculidae **Order:** Cuculiformes
Number of species: 127
Examples: cuckoo, common koel, roadrunner, smooth-billed ani, running coua

Waxbills

Family: Estrildidae **Order:** Passeriformes
Number of species: about 126
Examples: zebra finch, red avadavat, Java s parrow, blue-faced parrot-finch

American wood warblers

Family: Parulidae **Order:** Passeriformes
Number of species: about 125
Examples: bananaquit, yellow warbler, northern parula, painted redstart, ovenbird

Bulbuls

Family: Pycnonotidae **Order:** Passeriformes
Number of species: about 120
Examples: leaflove, red-whiskered bulbul, crested finchbill, bristlebill, nicator

Sunbirds

Family: Nectariniidae **Order:** Passeriformes
Number of species: about 116
Examples: ruby-cheeked sunbird, superb sunbird, long-billed spiderhunter

Starlings

Family: Sturnidae **Order:** Passeriformes
Number of species: 108
Examples: starling, common myna, yellow-billed oxpecker, metallic starling

Crows

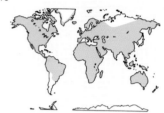

Family: Corvidae **Order:** Passeriformes
Number of species: 105
Examples: common crow, rook, magpie, blue jay, Hume's ground jay, piapiac

Icterids

Family: Icteridae **Order:** Passeriformes
Number of species: about 93
Examples: northern oriole, common grackle, Wagler oropendola, brown-headed cowbird

Gulls and terns

Family: Laridae **Order:** Charadriiformes
Number of species: about 90
Examples: herring gull, black-legged kittiwake, common tern, brown noddy

Kingfishers

Family: Alcedinidae **Order:** Coraciiformes
Number of species: about 86
Examples: kingfisher, belted kingfisher, laughing kookaburra

Sandpipers

Family: Scolopacidae **Order:** Charadriiformes
Number of species: about 82
Examples: whimbrel, American woodcock,
common snipe, ruff, red phalarope

Barbets

Family: Capitonidae **Order:** Piciformes
Number of species: about 81
Examples: double-toothed barbet,
coppersmith, yellow-fronted tinkerbird

White-eyes

Family: Zosteropidae **Order:** Passeriformes
Number of species: about 80
Examples: gray-breasted white-eye, Japanese
white-eye, Princip<\#142> island speirops

Swallows and martins

Family: Hirundinidae **Order:** Passeriformes
Number of species: about 78
Examples: barn swallow, blue rough-winged
swallow, house martin

Larks

Family: Alaudidae **Order:** Passeriformes
Number of species: about 75
Examples: skylark, horned lark, desert lark,
singing bushlark

Cotingas

Family: Cotingidae **Order:** Passeriformes
Number of species: 73
Examples: cock-of-the-rock, spangled cotinga,
bearded bellbird

Cuckoo-shrikes and minivets

Family: Campephagidae **Order:** Passeriformes
Number of species: about 70
Examples: large cuckoo-shrike, cicadabird,
white-winged triller, scarlet minivet

Nightjars

Family: Caprimulgidae **Order:**
Caprimulgiformes
Number of species: about 67
Examples: common poorwill, nightjar

Swifts

Family: Apodidae **Order:** Apodiformes
Number of species: about 67
Examples: common swift, brown needletail

Petrels

Family: Procellariidae **Order:**
Procellariiformes
Number of species: about 66
Examples: scaled petrel, northern fulmar

Shrikes

Family: Laniidae **Order:** Passeriformes
Number of species: about 65
Examples: northern shrike, gray-headed bush
shrike, black-backed puffback

Plovers

Family: Charadriidae **Order:** Charadriiformes
Number of species: about 63
Examples: lapwing, American golden plover,
wrybill, killdeer, common dotterel

Herons and bitterns

Family: Ardeidae **Order:** Ciconiiformes
Number of species: about 62
Examples: American bittern, black-crowned
night heron, gray heron, great egret

Wrens

Family: Troglodytidae **Order:** Passeriformes
Number of species: about 60
Examples: house wren, cactus wren, rock
wren, long-billed marsh wren

Falcons

Family: Falconidae **Order:** Falconiformes
Number of species: 60
Examples: caracara, peregrine falcon, kestrel,
hobby, gyrfalcon, collared falconet

Flowerpeckers

Family: Dicaeidae **Order:** Passeriformes
Number of species: about 58
Examples: crimson-breasted flowerpecker,
black berrypecker, mistletoe bird

Manakins

Family: Pipridae **Order:** Passeriformes
Number of species: about 56
Examples: blue-backed manakin

Lories

Family: Loriidae **Order:** Psittaciformes
Number of species: about 54
Examples: black-capped lory, dusky lory,
rainbow lorikeet

Pipits

Family: Motacillidae **Order:** Passeriformes
Number of species: about 54
Examples: golden pipit, water pipit, yellow
wagtail, yellow-throated longclaw

Woodcreepers

Family: Dendrocolaptidae **Order:**
Passeriformes
Number of species: about 50
Examples: barred woodcreeper, scythebill

Titmice

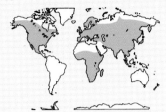

Family: Paridae **Order:** Passeriformes
Number of species: about 47
Examples: great tit, sultan tit, black-capped
chickadee

Tinamous

Family: Tinamidae **Order:** Tinamiformes
Number of species: about 47
Examples: great tinamou, solitary tinamou,
brown tinamou

Curassows

Family: Cracidae **Order:** Galliformes
Number of species: about 44
Examples: great curassow, nocturnal
curassow, crested guan, plain chachalaca

Hornbills

Family: Bucerotidae **Order:** Coraciiformes
Number of species: about 44
Examples: great Indian hornbill, helmeted
hornbill, southern ground hornbill

Vireos

Family: Vireonidae **Order:** Passeriformes
Number of species: about 43
Examples: red-eyed vireo, tawny-crowned
greenlet, chestnut-sided shrike-vireo

Birds of paradise

Family: Paradisaeidae **Order:** Passeriformes
Number of species: about 42
Examples: king bird of paradise, crinkle-
collared manucode, ribbon-tailed astrapia

Toucans

Family: Ramphastidae **Order:** Piciformes
Number of species: 37
Examples: toco toucan, emerald toucanet,
curl-crested aracari

Trogons

Family: Trogonidae **Order:** Trogoniformes
Number of species: about 35
Examples: quetzal, coppery-tailed trogon,
red-headed trogon

Puffbirds

Family: Bucconidae **Order:** Piciformes
Number of species: about 34
Examples: black-fronted nunbird, white-
necked puffbird

Mockingbirds

Family: Mimidae **Order:** Passeriformes
Number of species: about 31
Examples: mockingbird, gray catbird,
trembler, California thrasher

Threatened plants

The green kingdom of plants is threatened with decimation. Of the roughly 310,000 species so far identified, more than 6 percent are at risk. Destruction of the rain forests and other crucial habitats is the prime reason so many plants are in jeopardy. But pollution, modern agricultural practices, and urban development also consign species, if less spectacularly, to the threatened categories.

Species extinction is natural, but human intervention has accelerated the process more than a thousand times. And it is certain that species never identified are dying out, vanishing before they have ever been seen, let alone named. Even those known to us are becoming extinct before their uses are fully explored.

Wild relatives of crop plants such as rice and corn offer rich gene pools with the potential to improve yields or develop disease resistance. Some 40 percent of the world's drugs are derived from wild sources, suggesting that plants, especially from rain forests, represent a priceless storehouse of medicines. Yet other species, whether they are flowers, ferns, palms, or cycads, offer a wide assortment of horticultural prizes. Finally, plants provide resources valuable to industry, such as gums, waxes, fibers, timber, oils, or dyes.

The following list of threatened species, grouped in major geographical areas, gives an idea of the kind of plants the world is likely to lose if threats to natural environments continue at their present high level. The data for each species include its range, the agent causing its demise, its threatened status as categorized by the World Conservation Monitoring Centre, and (where known) the value the plant holds for people. The categories are defined as:

*Endangered: species in danger of extinction and whose survival is unlikely if causal factors continue to operate.

*Vulnerable: species likely to become endangered in the near future if causal factors continue to operate.

*Rare: species with small world populations which, while not vulnerable or endangered, are at risk.

NORTH AMERICA

Monterey Cypress, *Cupressus macrocarpa* CENTRAL CALIFORNIA (USA) Threat: disease and sea erosion of the cliffs where it grows. Status: rare. Value: prized as an ornamental and for its timber; it is also used in hedges and windbreaks.

Eureka Dune Grass, *Swallenia alexandrea* CALIFORNIA (USA) Threat: off-road-vehicles, such as trail bikes and dune buggies. Status: vulnerable. Value: the only member of its genus and so a genetic resource; its spreading rhizomes stabilize sand and dune slopes; it is also an important food source for some local fauna.

Texas Wild Rice, *Zizania texana* TEXAS (USA) This aquatic grass is a close relative of annual wild rice (Z. aquatica), which is grown as a crop plant in North America. It is palatable, nutritious, a potential crop, a genetic resource, and an important shelter for waterfowl. Its population has declined rapidly since it was first described in 1933 and it can only be found along the upper reaches of the San Marcos River in south-central Texas. Main threats include commercial collecting, sewage and chemical pollution, water recreation, floating debris, and silting due to dams. Many of these threats have been reduced and the wild rice's decline is slowing down, though its status remains vulnerable.

North American Pawpaw, *Asimina rugelii* FLORIDA (USA) Threat: urbanization. Status: endangered. Value: a small shrub cultivated as a garden plant.

Virginia Round-leaf Birch, *Betula uber* VIRGINIA (USA) Threat: low population numbers, vandalism, and overcollecting. Status: endangered. Value: unknown.

Louisiana Quillwort, *Isoetes louisianensis* GEORGIA, LOUISIANA (USA) Threat: road construction and canalization, Status: endangered. Value: important in the diet of local duck populations.

Tennessee Green Pitcher Plant, *Sarracenia oreophila* ALABAMA, GEORGIA, TENNESSEE (USA) Threat: urbanization, mining, and overcollecting. Status: endangered or vulnerable. Value: unknown.

Florida Arrowroot, *Zamia floridana* FLORIDA, GEORGIA (USA) Threat: tourist pressure. Status: vulnerable. Value: this cycad is an ornamental plant with potential as a food source.

Agave arizonica, ARIZONA (USA) Threat: overcollecting, grazing by deer and other herbivores, and possibly the use of shoots for human consumption. Status: endangered. Value: a succulent of horticultural interest.

Bandera County Ancistrocactus, *Ancistrocactus tobuschii* TEXAS (USA) Threat: overcollecting. Status: endangered. Value: prized by horticulturalists.

Knowlton Cactus, *Pediocactus knowltonii* NEW MEXICO, COLORADO (USA) Threat: overcollecting and human disturbance (recreation). Status: endangered. Value: of horticultural interest.

American Yellowwood, *Cladrastris lutea* USA Threat: flooding by dams and exploitation of saplings for nursery stock Status: vulnerable. Value: a deciduous tree of ornamental use; its wood was formerly employed in making gunstocks, its roots are a source of yellow dye.

CARIBBEAN ISLANDS

Tree Cactus, *Cereus robinii* FLORIDA (USA), CUBA Threat: urbanization, tourist development, and overcollecting. Status: vulnerable. Value: prized as a landscape ornamental.

Bermuda Cedar, *Juniperus bermudiana* BERMUDA Threat: introduced pests (scale insects) and tourist pressure. Status: vulnerable. Value: the wood of this evergreen is important in furniture making.

Tibouchina, *Tibouchina chamaecistus* MARTINQUE Threat: flower picking and overcollecting. Status: vulnerable. Value: this twisted dwarf shrub is a prized ornamental.

Palma Corcho, *Microcycas calcocoma* PINAR DEL RIO (Cuba) Threat: loss of habitat, Status: endangered. Value: this cycad is a prized ornamental.

CENTRAL AND SOUTH AMERICA

Neogomesia Cactus, *Ariocarpus agavoides* MEXICO Threat: overcollecting. Status: endangered. Value: prized by horticulturalists.

Caoba, *Persea theobromifolia* ECUADOR The tree is a canopy species of mature lowland wet forest and is restricted to the Rio Palenque Biological Center along the western base of the Andes Mountains in central Ecuador. Formerly abundant, forest clearance for the purpose of logging, banana and oil palm plantations, and settlement have made the tree an endangered species; in 1978 its population had been reduced to some 12 individuals. The timber quality of this fast-growing tree has been recognized—it was once an important source of commercial "mahogany." As a close relative of the cultivated avocado (*P. americana*) it is a genetic source of such characteristics as blight resistance.

Star Cactus, *Astrophytum asterias* MEXICO Threat: land clearance for agriculture. Status: rare. Value: of potential interest to horticulturalists.

Michay Rojo, *Berberidopsis corallina* CHILE Threat: habitat clearance for eucalyptus plantations. Status: endangered. Value: unknown.

Bamboo Cycad, *Ceratozamia hildae* MEXICO Threat: overcollecting. Status: rare. Value: a popular cultivated plant.

Caiapia, *Dorstenia albertorum* BRAZIL Threat: forest clearance for agriculture. Status: endangered. Value: unknown.

Aechmea dichlamydea VENEZUELA, TRINIDAD AND TOBAGO Threat: road building, fire, logging, and settlements. Status: vulnerable. Value: a bromeliad of horticultural interest.

Cacaode Monte, *Herrania balaensis* ECUADOR Threat: deforestation and road building. Status: endangered. Value: unknown.

Snow Mimosa, *Mimosa lanuginosa* BRAZIL Threat: land development. Status: vulnerable or endangered. Value: a suitable plant for tropical and subtropical gardens.

Dicliptera dodsonii ECUADOR Threat: deforestation (banana and oil palm plantations). Status: endangered. Value: a vine of potential interest to horticulturalists.

Costa Rican Jatropha, *Jatropha costaricensis* COSTA RICA Threat: livestock overgrazing and timber cutting. Status: endangered. Value: unknown.

Chilean Wine Palm, *Jubaea chilensis* CHILE Threat: tree cutting and land clearance. Status: vulnerable. Value: prized as an ornamental and an important domestic and commercial resource: baskets are made from leaves; palm honey and palm wine from sap; edible oil and food from seeds; sweetmeats from fruits.

Palo de Sal, *Pelliciera rhizophorae* COLOMBIA, ECUADOR, PANAMA, COSTA RICA Threat: clearance and development of mangrove swamps. Status: vulnerable. Value: unknown.

Diploperennial Teosinte, *Zea diploperennis* MEXICO Threat: land clearance for agriculture. Status: vulnerable. Value: this corn, a wild relative and probably ancestor of the cultivated variety (*Z. mays*), is a potential genetic resource. If its perennial habit and adaptability to cold and damp could be bred into corn cultivars, the agriculture of many countries would be transformed.

Ariocarpus agavoides MEXICO Threat: overcollecting. Status: vulnerable. Value: a cactus of unusual shape and habit prized by horticulturalists.

SOUTH ATLANTIC ISLANDS

Old Father Live Forever, Pelargonium

cotyledonis ST HELENA Threat: overgrazing by introduced goats. Status: endangered. Value: a perennial woody herb with the ability to stay alive for months without soil or water (hence the common name).

St Helena Redwood, Trochetiopsis erythroxylon ST HELENA Threat: overgrazing by introduced goats and timber cutting. Status: endangered. Value: this tree has hard, red-brown timber, a bark used in tanning, and potential as an ornamental.

EUROPE

Sicilian Fir, *Abies nebrodensis* SICILY Threat: exploitation, fire, and livestock overgrazing. Status: endangered. Value: a source of wood for papermaking.

Dwarf Beet, *Beta nana* GREECE Threat: livestock overgrazing and pressure from tourists. Status: rare. Value: this hairless perennial is a potential source of genetic variation for crop plants in the beet family.

Onosma tornensis, CZECHOSLOVAKIA, HUNGARY Threat: overgrazing (sheep, pigs), industrial practices (cement works), and tourism. Status: vulnerable. Value: a flowering plant suitable for rock gardens.

Hooded Helleborine Orchid, *Cephalanthera cucullata* CRETE Threat: livestock overgrazing. Status: endangered. Value: prized by horticulturalists.

Sisymbrium cavanillesianum SPAIN Threat: agriculture and livestock overgrazing. Status: a potential source of medicinal drugs.

Daphne rodriguezii BALEARIC IS. (Spain) Threat: tourist pressure. Status: endangered. Value: this small, dense, evergreen shrub has potential for rock and alpine gardens.

Cretan Date Palm, *Phoenix theophrasti* CRETE This many-stemmed palm, unique to Europe, can be found only in a few sites on the island, especially at Vai on the northeastern coast. It is a close relative of the cultivated date palm (*P. dactylifera*) and a source of genes governing such characteristics as resistance to pests, diseases, and cold. The palm, whose status is vulnerable, is threatened mainly by tourism—cars, camping, and fires prevent regeneration by destroying seedlings. The palm needs a high water table, so the species is also threatened by drainage and the pumping of freshwater to urban areas.

Dianthus callizonia ROMANIA Threat: tourism and livestock overgrazing. Status: rare. Value: a perennial herb suitable for rock gardens.

Lilium rhodopaeum BULGARIA, GREECE Threat: overcollecting and tourism. Status: rare. Value: a lily of great horticultural merit.

Artemisia granatensis SPAIN Threat: overcollecting. Status: endangered. Value: a perennial used as an ingredient in "Artemisia Tea" and for alcoholic drinks.

Shrubby Cress-rocket, *Vella pseudocytisus* SPAIN Threat: overgrazing (sheep, goats), destruction of habitat, and overcollecting. Status: vulnerable. Value: a leafy shrub, once

cultivated and used for fuel, and a potent ornamental.

Caralluma distincta KENYA, TANZANIA Threat: agriculture and livestock

Wild Rhubarb, *Rheum rhaponticum* BULGARIA, NORWAY Threat: uprooting. Status: rare. Value: a robust perennial economically and historically important as a parent of the cultivated rhubarb (*R. X cultorum*); it is a potential genetic resource and has medicinal and culinary uses.

Lundy Cabbage, *Rhynchosinapsis wrightii* LUNDY IS. (England) Threat: overgrazing (goats, deer) and competition with bracken. Status: rare. Value: a close relative of the cultivated cabbages (Brassica spp.), and a valuable genetic resource.

Doronicum cataractarum AUSTRIA Threat: competition with other plants, grazing by cattle, and overcollecting. Status: rare. Value: an alpine perennial suitable for rock gardens.

AFRICA

Spiral Aloe, *Aloe polyphylla* LESOTHO Threat: overcollecting. Status: vulnerable. Value: a prized garden succulent.

Yeheb Nut, *Cordeauxia edulis* ETHIOPIA, SOMALIA Threat: livestock overgrazing, shifting cultivation, warfare. Status: endangered. Value: the bush provides dyes and the nuts are a potential source of food.

Tarout Cypress, *Cupressus dupreziana* ALGERIA The tree was a major local source of timber in the 19th century, but by 1978 its population had been reduced to 153 living specimens. This decline had been compounded by nomads sheltering beneath the trees; they burned the wood and their livestock ate the seedlings. The tiny remaining population survives in an area of 77sq miles in the Tassili N'Ajjer mountain range in southeast Algeria. The species is still endangered despite attempts to improve its status. The tarout is an important source of genes, such as those for frost and drought tolerance, which can be bred into other cypresses. Its medium density, aromatic wood has commercial value, and the tree rings of dead specimens (some of which are around 2,000 years old) could provide vital clues to the Sahara Desert's past climate.

Wild Olive, *Olea laperrinei* SOUTHERN ALGERIA, NIGER, SUDAN Threat: livestock overgrazing, woodcutting, and environmental changes. Status: vulnerable. Value: a close relative of the cultivated olive (*O. europaea*), and a valuable genetic resource.

Nubian Dragon Tree, *Dracaena ombet* DJIBOUTI, ETHIOPIA, SOMALIA, SUDAN Threat: cutting for firewood, drought, and livestock overgrazing. Status: vulnerable or rare. Value: the leaves are used for weaving mats and baskets.

Euphorbia cameronii SOMALIA Threat: livestock overgrazing, warfare, and environmental changes. Status: endangered. Value: this succulent shrub has economic potential as livestock fodder; it is a source of moisture for grazing animals in dry periods, and may be used as an ornamental.

Caralluma distincta KENYA, TANZANIA Threat: agriculture and livestock overgrazing. Status: endangered. Value: a succulent herb of potential horticultural interest; it may contain pharmacologically active substances, as do other members of the Asclepiadaceae (milkweed) family, to which it belongs.

Centaurea junoniana CANARY IS. Threat: volcanic eruptions. Status: endangered. Value: woody perennial of horticultural interest.

Aeonium nobile CANARY IS. Threat: agriculture, grazing by goats, and overcollecting. Status: vulnerable. Value: a succulent herb of horticultural potential.

Madrono, *Arbutus canariensis* CANARY IS. Threat: loss of laurel forest habitat (agriculture, fuel wood). Status: vulnerable. Value: a tree of horticultural interest; its edible fruit is rich in vitamin C and is reputed to be the Golden Apple of the Hesperides in Greek mythology.

Tedera Salvaje, *Ruta pinnata* CANARY IS. Threat: clearing of vegetation (agriculture, livestock overgrazing). Status: vulnerable. Value: a relative of the common rue (*R. graveolens*) and rich in oil glands which produce potentially useful chemicals.

Pininana, *Echium pininana* CANARY IS. Threat: habitat clearance for agriculture, livestock overgrazing, and overcollection. Status: endangered. Value: this giant herb is a spectacular garden plant suitable for mediterranean climates.

Dragon Blood Tree, Dracaena draco CANARY IS., CAPE VERDE IS., and MADEIRA Threat: commercial exploitation for its sap. Status: vulnerable. Value: the sap is used to produce red resin for varnishes, paper pigments, and pharmaceutical plasters, and may also have medicinal properties.

Papyrus, *Cyperus papyrus hadidi* EGYPT The habitat of this giant perennial sedge is restricted to the Wadi Natroun depression, west of the southernmost part of the Nile Delta. The papyrus is threatened by the extraction of water from the Nile, accompanied by changes in irrigation patterns, which have caused freshwater ponds and swamps to dry out and become more salty. In Egypt its status is endangered. The papyrus was the symbol of the ancient Kingdom of Lower Egypt, where it was used for making sandals, mats, and paper, and as a source of medicines.

Jasmine-flowered Heath, *Erica jasminifolia* SOUTH AFRICA Threat: habitat clearance (agriculture, road construction, and fire). Status: endangered. Value: this evergreen shrub has potential as a garden plant.

Golden Gladiolus, *Gladiolus aurea* SOUTH AFRICA Threat: competition from introduced plants, flower picking, and modified land drainage due to gravel extraction. Status: endangered. Value: this slender herbaceous plant has horticultural potential.

Dalla, *Medemia argun* EGYPT, SUDAN Threat: habitat clearance for agriculture and by irrigation schemes. Status: endangered. Value: this palm, the only member of its genus, is a potential genetic resource.

African Violet, *Saintpaulia ionantha* TANZANIA Threat: forest clearance and habitat disturbance. Status: endangered. Value: one of the world's most popular perennial houseplants.

Flor de Mayo, *Senecio hadrosomus* CANARY IS. Threat: livestock overgrazing and increasing aridity of the climate. Status: endangered. Value: a wild relative of florist's cineraria and a potential genetic resource for horticulturalists.

INDIAN OCEAN ISLANDS

Catharanthus coriaceus MADAGASCAR Threat: livestock overgrazing and fire. Status: endangered. Value: a relative of the rosy periwinkle (*C. roseus*), which is used to treat leukemia, and potentially of great pharmacological value.

Mauritian Crinum Lily, Crinum mauritianum MAURITIUS Threat: dam and reservoir construction. Status: endangered. Value: a large bulbous plant prized by horticulturalists.

Bois de Prune Blanc, *Drypetes caustica* MAURITIUS Threat: tree felling and low breeding potential. Status: endangered. Value: this tree provides good hard timber, and its mustard oils have potential chemical value.

Cucumber Tree, *Dendrosicyos socotranus* SOCOTRA IS, (possibly SOUTH YEMEN) Threat: overgrazing by goats and overcollecting for camel fodder. Status: vulnerable. Value: this giant obese succulent is a botanical curiosity and the only arborescent member of the cucumber family.

Saiberbher, *Begonia socotrana* SOCOTRA IS. (possibly SOUTH YEMEN) Threat: livestock overgrazing. Status: endangered. Value: this begonia is an important genetic resource, able to endow begonia hybrids with the capacity for winter-flowering.

Dirachma, *Dirachma socotrana* SOCOTRA IS. (possibly SOUTH YEMEN) Threat: livestock overgrazing. Status: endangered. Value: the only member of its family, it is a potential genetic resource.

Coco de Mer, *Lodoicea maldivica* SEYCHELLES Threat: overcollecting of nuts for tourists, habitat clearance, fire, and competition from introduced plants. Status: vulnerable. Value: this palm produces the biggest seeds in the world (up to 66lb); these are considered to have aphrodisiac properties.

Toxocarpus schimperianus SEYCHELLES Threat: habitat loss (fire and forest exploitation cause erosion, loss of soil fertility, and drought). Status: vulnerable. Value: the plant may contain pharmacologically active substances, as do other members of the Asclepiadaceae (milkweed) family, to which it belongs.

Bois de Fer, *Vateria seychellarum* SEYCHELLES Threat: habitat loss (fire and forest exploitation cause erosion, loss of soil fertility, and drought). Status: endangered. Value: the tree provides good timber and is a valuable genetic resource.

Big-leaf Palm, *Marojeja darianii* MADAGASCAR Threat: overcollecting for food and habitat clearance for agriculture. Status: endangered. Value: unknown.

Socotran Pomegranate, *Punica protopunica* SOCOTRA IS. (possibly SOUTH YEMEN) In 1953 this small tree was the chief constituent of thickets on Socotra's limestone slopes, but by 1978 the species had become endangered, with only four old specimens surviving. The rapid decline was caused by heavy overgrazing by introduced livestock, especially goats and cattle. A relative of the cultivated pomegranate (*P. granatum*), it is an important genetic resource; its palatable fruit is eaten by the local Bedu people. The tree is one of 216 flowering plants endemic to Socotra and the neighboring island of Abd al Kuri; almost half of these are seriously endangered.

ASIA

Grantham's Camellia, *Camellia granthamiana* HONG KONG, CHINA Threat: naturally low population numbers. Status: endangered. Value: a small tree, favored as a genetic resource for the breeding of new garden cultivars.

Coptis, *Coptis teeta* BURMA, CHINA, INDIA Threat: overexploitation. Status: vulnerable. Value: an herb widely used in local medicine and therefore a potential source of pharmaceutical drugs.

Dawn Redwood, *Metasequoia glyptostroboides* SICHUAN PROVINCE (China) Threat: agriculture and the cropping of seedlings. Status: rare. Value: this tree, a "living fossil," is of great botanical and horticultural interest.

Maxburretia rupicola PENINSULAR MALAYSIA Threat: overcollecting, quarrying, and fires. Status: rare. Value: the palm's prized flowers are of horticultural interest.

Palms (*Family* **Palmae**) Most of the roughly 2,600 species of palm are native to tropical regions, especially to Asia and Central/South America, and to a lesser extent Africa. Some 83 percent are found only in the wild, while the remaining 17 percent exist both in cultivation and in the wild. Their usefulness to man has a long history. In order of importance their uses are: food, fiber, building materials, fuel wood, and folk medicines. No palm species has been confirmed as totally extinct, but 91 are said to be endangered (45 in the New World and 46 in the Old World). They are threatened mainly by forest clearance for agriculture, settlements, the creation of pastures, mining, and hydroelectric power operations.

Musa gracilis PENINSULAR MALAYSIA Threat: habitat clearance for agriculture. Status: vulnerable. Value: this giant relative of the banana is of botanical interest and a potential genetic resource.

Low's Pitcher Plant, Nepenthes lowii BORNEO Threat: overcollecting. Status: rare. Value: this insectivorous plant is a horticultural prize.

Snow Orchid, *Diplomeris hirsuta* WEST BENGAL (India) Threat: livestock overgrazing and landslips. Status: vulnerable. Value: of great horticultural merit; tubers may contain useful alkaloids.

Drury's Slipper Orchid, *Paphiopedilum druryi* INDIA Threat: overcollecting and forest fires. Status: endangered, possibly extinct. Value: a horticultural gem and a possible source of alkaloids and medicinal drugs.

Giant Rafflesia, *Rafflesia arnoldii* SUMATRA Threat: logging, shifting cultivation, and overcollecting. Status: vulnerable. Value: this parasitic plant has the world's largest flowers; locally it is a source of medicines and may be of pharmaceutical interest.

Jade Vine, *Strongylodon macrobotrys* PHILIPPINES Threat: deforestation. Status: vulnerable. Value: a spectacular plant in cultivation.

Wallich's Elm, *Ulmus wallichiana* AFGHANISTAN, INDIA, NEPAL, PAKISTAN Threat: cutting for animal fodder. Status: endangered. Value: the fibrous bark is used to make rope and the wood could be a source of timber.

Blue Vanda of Asia, *Vanda caerulea* BURMA, INDIA, NEPAL Threat: overcollecting. Status: vulnerable. Value: an epiphytic orchid prized by horticulturalists.

AUSTRALIA AND NEW ZEALAND

Waddy, *Acacia peuce* AUSTRALIA Threat: timber cutting and overgrazing by cattle. Status: vulnerable. Value: a tree valued for its hard, durable wood and its ability to grow in extreme desert conditions.

Mogumber Bell, *Darwinia carnea* SOUTHWEST AUSTRALIA Threat: overgrazing by sheep and rabbits. Status: endangered. Value: a sought-after garden plant.

Augusta Kennedia, *Kennedia macrophylla* SOUTHWEST AUSTRALIA Threat: urbanization, road works, and recreation. Status: endangered. Value: a member of the pea family, it is a prized garden plant and a potential genetic resource.

Underground Orchid, *Rhizanthella gardneri* SOUTHWEST AUSTRALIA Threat: farming and agriculture. Status: endangered. Value: one of only two underground orchids in the world (the other is *Cryptanthemis slateri* from NSW and Queensland), it is a popular plant with great interest for botanists.

Byfield Fern, *Bowenia serrulata* AUSTRALIA Threat: overcollecting and destruction because toxic to livestock. Status: vulnerable. Value: a distinctive cycad prized as an ornamental by horticulturalists.

Cycads (*Order* **Cycadales**)

Said to be living fossils because they were thriving at a time when dinosaurs roamed the world (between 150 and 200 million years ago). Cycads are long-lived, palmlike trees that grow and reproduce at a slow rate. They are threatened largely by overcollecting for the horticultural trade or by habitat clearance for agriculture, the creation of pastures, settlements, roads, dams, and tourist developments. They are among the most endangered of plant groups: of the 168 species, 14 are thought to be endangered, 40 vulnerable, 32 rare, 6 indeterminate, and one extinct. The remainder are either not threatened or else their status is insufficiently documented.

Horseshoe Fern, *Marattia salicina* LORD HOWE IS. (Australia) Threat: overcollecting and grazing by feral pigs. Status: endangered. Value: prized by horticulturalists.

Lobster Claw, *Clianthus puniceus* NORTH IS. (New Zealand) Threat: overgrazing by introduced mammals (Australian opossums, goats, deer, pigs). Status: endangered. Value: this woody shrub is an outstanding horticultural plant.

Three Kings Cabbage Tree, *Cordyline kaspar* THREE KINGS IS. (New Zealand) Threat: low population numbers following overgrazing by introduced mammals (now removed). Status: rare. Value: a small tree useful for planting in coastal districts.

Tecomanthe speciosa THREE KINGS IS. (New Zealand) Threat: low population due to past degradation of habitat (humans, goats). Status: endangered. Value: an evergreen vine of potential interest to horticulturalists.

White Gum, *Eucalyptus argophloia* AUSTRALIA Threat: forest clearance for agriculture. Status: endangered. Value: this tree has hard timber, can act as a windbreak or shade, and makes a good cultivated plant.

Hamilton's Gunnera, *Gunnera hamiltonii* STEWART IS. (New Zealand) Threat: quarrying, competition from introduced plants (weeds, duneland stabilizers), and pressure from livestock overgrazing. Status: endangered. Value: a small stout plant attractive in a rock garden.

Chatham Island's Forget-me-not, *Myosotidium hortensia* CHATHAM IS. (New Zealand) Threat: trampling and grazing by introduced mammals (pigs, sheep). Status: endangered. Value: a succulent perennial prized as a garden plant.

PACIFIC OCEAN ISLANDS

Kauai Silversword, *Argyroxiphium kauense* HAWAII Threat: overgrazing by sheep. Status: endangered. Value: unknown.

Kauai Hesperomannia, *Hesperomannia lydgatei* HAWAII Threat: deforestation. Status: endangered. Value: unknown.

Neowawraea phyllanthoides HAWAII Threat: low population due to past overgrazing (cattle, goats) and rooting by pigs. Status: endangered. Value: a tree with hard, heavy, close-grained timber that is potentially good for cabinet-making.

Hayun Lago, *Serianthes nelsonii* GUAM, ROTA (Western Pacific) Threat: grazing of seedlings by introduced deer and land development. Status: endangered. Value: a tree with good timber and potential as an ornamental.

Chonta, *Juania australis* JUAN FERNANDEZ IS. Threat: overgrazing, cutting of wood for souvenirs. Status: rare. Value: this palm is the only member of its genus and is therefore a genetic resource; the "heart" of the wood can be eaten and the wood itself is used for walking sticks and furniture.

Miconia, *Miconia robinsoniana* GALAPAGOS IS. Threat: overgrazing by introduced mammals and competition from introduced plants. Status: vulnerable. Value: unknown.

Endangered species: mammals

The survival of an increasing number of animal species is at risk, largely because mankind is destroying their habitats. In 1988, the total of known threatened species stood at 4,589. This comprised 555 mammals, 1,073 birds, 186 reptiles, 54 amphibians, 596 fish, and 2,125 invertebrates. Estimates suggest that if present trends continue, several hundred vertebrates and roughly a million insects will become extinct by 2040.

The following checklists are composed of endangered animals onl—they do not contain species classed as vulnerable, rare, indeterminate, or insufficiently known. The World Conservation Monitoring Centre defines an endangered species as "one in danger of extinction and whose survival is unlikely if the causal factors continue operating."

Relatively few groups of animal species have had their status thoroughly assessed. The following lists therefore represent only a selection of the most endangered species, with the mammals being the most comprehensive and the invertebrates, about which least is known, the least. Data for each species include its common and scientific names, distribution, and the main agents of threat. Profiles give added detail to the plight of certain species and, where known, state the CITES appendix to which they have been assigned.

CITES (the Convention of International Trade in Endangered Species of Wild Fauna and Flora) has three appendices: I contains species that may be threatened with extinction by trade; II contains species likely to become threatened with extinction if trade continues; III is little used and contains species protected nationally but not internationally.

Order
MARSUPIALIA (Marsupials)

Brush-tailed Bettong, *Bettongia penicillata* AUSTRALIA Threat: habitat losses (seasonal burning) and predation by introduced mammals (foxes).

Woodlark Island Cuscus, *Phalanger lullulae* WOODLARK IS. (Papua New Guinea) Threat: hunting and habitat loss (commercial logging).

Northern Hairy-nosed Wombat, *Lasiorhinus kreftii* AUSTRALIA Threat: competition with rabbits and cattle, disease, and drought.

Greater Bilby, *Macrotis lagotis* NORTHERN TERRITORY (Australia) Threat: human disturbance, occasional hunting, and competition with rabbits.

Numbat, *Myrmecobius fasciatus* AUSTRALIA Threat: land clearance (agriculture), brush fires, and predation by introduced mammals (foxes, cats, dogs).

Brindled Nailtail Wallaby, *Onychogalea fraenata* QUEENSLAND (Australia) Once common in New South Wales and Queensland, it was thought to be extinct until 1974, when a viable population was discovered north of Dingo in central Queensland. Its scrub woodland and tall shrubland habitat is threatened by large-scale pasture development for beef cattle. The wallaby is also threatened by introduced herbivores (rabbits) and predators (foxes). Listed in CITES Appendix I.

Order
INSECTIVORA (Insectivores)

Cuban Solenodon, *Solenodon cubanus* CUBA Threat: deforestation.

Haltian Solenodon, *Solenodon paradoxus* HISPANIOLA Threat: deforestation, land development, and predation (dogs).

Giant Golden Mole, *Chrysospalax trevelyani* SOUTH AFRICA Threat: uncertain, possibly drought and habitat loss.

Order
CHIROPTERA (Bats)

Chapman Fruit Bat, *Dobsonia exoleta chapmani* NEGROS IS. (Philippines) Threat: deforestation and hunting (food). Possibly extinct in the wild.

Philippines Tube-nosed Fruit Bat, *Nyctimene rabori* NEGROS IS. (Philippines) Threat: deforestation.

Comoro Black Flying Fox, *Pteropus livingstonii* COMORO IS. (Indian Ocean) Threat: uncertain.

Rodrigues Flying Fox, *Pteropus rodricensis* RODRIGUES (Mauritius) Threat: starvation (loss of fruit food in habitat), hunting, and cyclones.

Samoan Flying Fox, *Pteropus samoensis* SAMOA, FIJI IS. Threat: uncertain.

Seychelles Sheath-tailed Bat, *Coleura seychellensis* SEYCHELLES Threat: uncertain.

Puerto Rican Flower Bat, *Phyllonycteris major* PUERTO RICO Threat: uncertain. Possibly extinct in the wild.

Tanzanian Woolly Bat, *Kerivoula africana* TANZANIA Threat: uncertain. Possibly extinct in the wild.

Gray Bat, *Myotis grisescens* SOUTHEAST USA The bat inhabits isolated limestone caves and by the mid-1970s had declined to around 1.5 million, one-fifth of the estimated population 20 years previously. In winter, nine caves harbor about 95 percent of the total population and more than half occupy a single cave. The species is threatened by cave disturbance and commercialization, wanton killing by vandals, flooding caused by dams and reservoirs, siltation from open-cast mining; water pollution, and pesticides. Protection of its cave habitats has led to stabilization or increase in populations in the 1980s. Listed as endangered on the USA Endangered Species List.

Order
PRIMATES (Monkeys and apes)

Broad-nosed Gentle Lemur, *Hapalemur simus* MADAGASCAR Threat: deforestation.

Sclater Lemur, *Lemur macaco flavifrons* MADAGASCAR Threat: deforestation.

Hairy-eared Dwarf Lemur, *Allocebus trichotis* MADAGASCAR Threat: uncertain (human interference thought unlikely).

Indris, *Indri indri* MADAGASCAR Threat: deforestation (logging, agriculture).

Aye-aye, *Daubentonia madagascariensis* MADAGASCAR Threat: deforestation and persecution.

Philippine Tarsier, *Tarsius syrichta* PHILIPPINES Threat: deforestation and capture (export trade).

Woolly Spider Monkey, *Brachyteles arachnoides* SOUTHEAST BRAZIL Threat: deforestation (fuel, settlement, agriculture), and hunting (food).

Southern Bearded Saki, *Chiropotes satanas satanas* BRAZIL Threat: deforestation (settlement).

Yellow-tailed Woolly Monkey, *Lagothrix flavicauda* ECUADOR, PERU Threat: hunting (food, skin), capture (export trade), and deforestation (pasture lands, agriculture).

Central American Squirrel Monkey, *Saimiri oerstedi* COSTA RICA, PANAMA Threat: deforestation (banana and sugarcane plantations, cattle farms) and capture (export trade).

Buffy-tufted-ear Marmoset, *Callithrix aurita* SOUTHEAST BRAZIL Threat: deforestation.

Buffy-headed Marmoset, *Callithrix flaviceps* SOUTHEAST BRAZIL Threat: deforestation (settlement) and possibly capture (scientific research, pets).

Golden Lion Tamarin, *Leontopithecus rosalia* SOUTHEAST BRAZIL Threat: deforestation (agriculture, pasturelands, urban development).

Golden-headed Lion Tamarin, *Leontopithecus chrysomelas* EAST BRAZIL Threat: deforestation (cocoa plantations).

Golden-rumped Lion Tamarin, *Leontopithecus chrysopygus* BRAZIL Threat: deforestation.

Cotton-top Tamarin, *Saguinus oedipus oedipus* NORTHWEST COLOMBIA Threat: deforestation (agriculture, pasturelands) and capture (pet trade, biomedical research).

Tana River Mangabey, *Cercocebus galeritus galeritus* KENYA Threat: deforestation (agriculture, seasonal flooding, burning).

White-throated Guenon, *Cercopithecus erythrogaster* BENIN, NIGERIA Threat: deforestation (commercial logging, agriculture), road building, mining/quarrying, and hunting (food).

Russet-eared Guenon, *Cercopithecus erythrotis* CAMEROON, BIOKO, NIGERIA Threat: deforestation and hunting (food).

Preuss Guenon, *Cercopithecus preussi* CAMEROON, BIOKO, NIGERIA) Threat: deforestation (commercial logging) and hunting (food).

Black Colobus, *Colobus satanas* CAMEROON, CONGO, EQUATORIAL GUINEA, GABON Threat: deforestation (commercial logging) and hunting (food, skin).

Red Colobus, *Colobus badius* [subspecies] GHANA, IVORY COAST, CONGO, BIOKO, CAMEROON, KENYA, TANZANIA Threat: habitat loss.

Kirk Colobus, *Colobus kirki* ZANZIBAR Threat: habitat degradation (woodcutting, human encroachment).

Drill, *Mandrillus leucophaeus* CAMEROON, BIOKO, NIGERIA The drill inhabits lowland and evergreen rain forest, usually in hilly, rock-strewn terrain by the coast. One of the most endangered monkeys in Africa, it is threatened mainly by the deforestation caused by commercial logging and the establishment of plantations; it is also hunted for food and as a crop pest. Listed in CITES Appendix I.

Mentawai Macaque, *Macaca pagensis* MENTAWAI IS. (Indonesia) Threat: deforestation, hunting (food), and capture (export trade).

Lion-tailed Macque, *Macaca silenus* SOUTH INDIA Threat: deforestation (plantations, hydroelectric power), hunting (food), and capture (export trade).

Javan Leaf Monkey, *Presbytis comata* JAVA Threat: loss of forest habitat (settlement).

Mentawai Leaf Monkey, *Presbytis potenziani* MENTAWAI IS. (Indonesia) Threat: deforestation (logging) and hunting (food).

Tonkin Leaf Monkey, *Presbytis francoisi* INDOCHINA, SOUTHWEST CHINA Threat: loss of forest habitat (deforestation, warfare) and hunting (medicines).

Nilgiri Leaf Monkey, *Presbytis johni* SOUTH INDIA Threat: deforestation (agriculture, pasturelands, wood, plantations) and hunting (food).

White-headed Black Leaf Monkey, *Trachypithecus leucocephalus* SOUTH CHINA Threat: loss of forest habitat and hunting (medicines).

Red-shanked Douc Monkey, *Pygathrix nemaeus* CENTRAL VIETNAM, LAOS Threat: deforestation (warfare, defoliation) and hunting (food).

Black-shanked Douc Monkey, *Pygathrix nigripes* SOUTHERN VIETNAM, SOUTHERN LAOS, EASTERN KAMPUCHEA Threat: loss of forest habitat (warfare, agriculture) and hunting (food).

Tonkin Snub-nosed Monkey, *Pygathrix avunculus* NORTHERN VIETNAM Threat: deforestation (warfare, agriculture) and hunting (food).

Yunnan Snub-nosed Monkey, *Pygathrix roxellana* CHINA Threat: hunting (skin). Population very low—around 200.

Ghizhou Snub-nosed Monkey, *Pygathrix brelichi* CHINA Threat: habitat loss and hunting.

Pig-tailed Langur, *Simias concolor* MENTAWAI IS. (Indonesia) Threat: deforestation (commercial logging) and hunting (food).

Kloss Gibbon, *Hylobates klossi* MENTAWAI IS. (Indonesia) Threat: deforestation (commercial logging) and hunting (food).

Javan Gibbon, *Hylobates moloch* JAVA Threat: deforestation (commercial logging) and capture (export trade).

Mountain Gorilla, *Gorilla gorilla beringei* RWANDA, UGANDA, ZAIRE Threat: deforestation, human disturbance, and encroachment.

Eastern Lowland Gorilla, *Gorilla gorilla graueri* EASTERN ZAIRE Threat: deforestation, human disturbance, and encroachment.

West African Chimpanzee, *Pan trogolodytes verus* WEST AFRICA Threat: habitat loss (agriculture) and capture (export trade).

Orangutan, *Pongo pygmaeus* BORNEO, SUMATRA Threat: deforestation (commercial logging, agriculture) and capture (export trade).

Order
EDENTATA (Edentates)

Maned Sloth, *Bradypus torquatus* SOUTHWEST BRAZIL Threat: deforestation.

Order
LAGOMORPHA (Rabbits and hares)

Riverine Rabbit, *Bunolagus monticularis* SOUTH AFRICA Threat: habitat loss (agriculture).

Hispid Hare, *Caprolagus hispidus* FOOTHILLS of the HIMALAYAS (Asia) Threat: habitat loss (forestry, dry season burning, agriculture, settlement) and hunting (food).

Tehuantepec Hare, *Lepus flavigularis* MEXICO Threat: uncertain.

Amami Rabbit, *Pentalagus furnessi* RYUKI IS. (Japan) Threat: deforestation (settlement) and predation (feral dogs).

Volcano Rabbit, *Romerolagus diazi* MEXICO The rabbit inhabits scattered pines and tussocky grasses on the upper-middle slopes of two volcanoes, Popocatepetl and Ixtacihuatl, and several adjacent mountains. It is threatened by the loss of its habitat and wanton shooting. Listed in CITES Appendix I.

Order
RODENTIA (Rodents)

Vancouver Island Marmot, *Marmota vancouverensis* CANADA Threat: deforestation (commercial logging, development of ski resorts).

Delmarva Fox Squirrel, *Sciurus niger cinereus* MARYLAND (USA) Threat: deforestation (logging).

San Quintin Kangaroo Rat, *Dipodomys gravipes* BAJA CALIFORNIA (Mexico) Threat: habitat loss (urbanization, development, agriculture).

Morro Bay Kangaroo Rat, *Dipodomys heermanni morroensis* CALIFORNIA (USA) Threat: habitat loss (changes in vegetation, urbanization).

Fresno Kangaroo Rat, *Dipodomys nitratoides exilis* FRESNO COUNTY, CALIFORNIA (USA) Threat: habitat loss (agriculture, development, urbanization).

Tipton Kangaroo Rat, *Dipodomys nitratoides nitratoides* CALIFORNIA (USA) Threat: habitat loss (agriculture, urbanization, development).

Salinas Pocket Mouse, *Perognathus inornatus psammophilus* CALIFORNIA (USA) Threat: habitat loss (agriculture, development, urbanization).

Los Angeles Pocket Mouse, *Perognathus longimembris brevinasus* CALIFORNIA (USA) Threat: habitat loss (urbanization, development).

Saltmarsh Harvest Mouse, *Reithrodontomys raviventris* CALIFORNIA (USA) Threat:

habitat loss (urban and industrial development, diking for salt ponds, water pollution).

Colorado River Cotton Rat, *Sigmodon arizonae plenus* CALIFORNIA (USA) Threat: habitat loss (agriculture, urbanization).

Amargosa Vole, *Microtus californicus scirpensis* CALIFORNIA (USA) Threat: habitat loss (agriculture, urbanization).

Cabrera Hutia, *Capromys angelcabrerai*; **Large-eared Hutia**, *Capromys auritus*; **Little Earth Hutia**, *Capromys snafelipensis* CUBA Threat: hunting by local fishermen, human disturbance, and a naturally low population.

Dwarf Hutia, *Capromys nanus* CUBA Threat: habitat loss (agriculture).

Order
CETACEA (Whales)

Indus River Dolphin, *Platanista minor* PAKISTAN Threat: habitat loss (impoundment of water and its withdrawal for irrigation) and exploitation by local fishermen.

Yangtze River Dolphin, *Lipotes vexillifer* YANGTZE RIVER (China) Threat: incidental capture in fishing nets and collision with boats; decrease in food supply (overfishing).

Blue Whale, *Balaenoptera musculus* ALL OCEANS Threat: overhunting.

Humpback Whale, *Megaptera novaeangliae* ALL OCEANS. There are thought to be two separate populations, northern and southern: the former was once some 15,000 strong and the latter 100,000. Whaling in the 19th century and the first half of the 20th severely reduced these populations. By the 1980s, the world total was thought to be less than 7,000. Despite a moratorium on whaling, the species is still threatened by hunters and by incidental capture in fishing nets. Listed in CITES Appendix I.

Bowhead Whale, *Balaena mysticetus* ARCTIC, NORTH ATLANTIC, and NORTH PACIFIC OCEANS Threat: overhunting.

Northern Right Whale, *Eubalaena glacialis* NORTH ATLANTIC and NORTH PACIFIC OCEANS Threat: overhunting.

Order
CARNIVORA (Carnivores)

Red Wolf, *Canis rufus* TEXAS, LOUISIANA (USA) Threat: habitat loss, trapping, poisoning, and hybridization with coyotes.

Simien Fox, *Canis simensis* ETHIOPIA Threat: habitat loss (agriculture) and hunting.

Baluchistan Bear, *Selenarctos thibetanus gedrosianus* IRAN, PAKISTAN Threat: persecution by farmers.

Barbados Raccoon, *Procyon gloveralleni* BARBADOS Threat: hunting and possible competition from other raccoon species. Possibly extinct in the wild.

Black-footed Ferret, *Mustela nigripes* USA Threat: habitat loss (agriculture, pasture), disease (distemper), poisoning, predation, and occasional hunting.

Malabar Large Spotted Civet, *Viverra megaspila civettina* WESTERN GHATS (India) Threat: uncertain, probably persecution and loss of habitat to agriculture.

Sokoke Bushy-tailed Mongoose, *Bdeogale crassicauda omnivora* NORTH TANZANIA, KENYA Threat: uncertain.

Liberian Mongoose, *Liberiictis kuhni* NORTHEAST LIBERIA (possibly IVORY COAST, GUINEA) Threat: habitat loss, hunting (food).

Barbary Hyena, *Hyaena hyaena barbara* ALGERIA, MOROCCO, TUNISIA Threat: habitat loss (settlement, agriculture).

Asiatic Cheetah, *Acinomyx jubatus venaticus* IRAN, USSR Threat: habitat loss (agriculture), and hunting (fur and as a stork predator).

Florida Cougar, *Felis concolor coryi* USA Threat: hunting, habitat loss, road kills, and reduced prey numbers.

Eastern Cougar, *Felis concolor cougar* CANADA, USA Threat: hunting, reduced prey numbers, and habitat loss.

Iriomote Cat, *Felis iriomotensis* RYUKU IS. (Japan) Threat: deforestation (agriculture).

Pakistan Sand Cat, *Felis margarita scheffeli* PAKISTAN Threat: capture (export trade).

Pardel Lynx, *Felis pardina* PORTUGAL, SPAIN Threat: habitat loss, disease-reduced prey numbers, and hunting.

Asiatic Lion, *Panthera leo persica* GUJERAT (India) Threat: habitat loss (overgrazing by cattle, water buffalo).

Tiger, *Panthera tigris* ASIA Threat: habitat loss (commercial logging, tree-planting, farming, settlement) and hunting (sport, skin, persecution).

Snow Leopard, *Panthera uncia* ASIA Threat: hunting (skin, possible persecution).

Japanese Sea Lion, *Zalophus californianus japonicus* JAPAN, NORTH and SOUTH KOREA Threat: persecution by local fishermen, human disturbance, and possible reduction of fish prey.

Mediterranean Monk Seal, *Monachus monachus* MEDITERRANEAN and MAURITIAN COASTS Threat: persecution by local fishermen, reduction of fish prey, incidental drowning in fishing nets, pollution of the sea, and human disturbance.

Hawaiian Monk Seal, *Monachus schauinslandi* HAWAII Threat: past uncontrolled killing, and recent disturbance from men, dogs, and sharks.

Caribbean Monk Seal, *Monachus tropicalis* CARIBBEAN Threat: past uncontrolled killing and recent human disturbance. Probably extinct.

Saimaa Seal, *Phoca hispida saimensis* FINLAND Threat: persecution by local fishermen and possible pollution.

Order
PROBOSCIDEA (Elephants)

Asian Elephant, *Elephas maximurs* ASIA In the past these elephants inhabited regions from Syria to Southeast Asia and were common on the Indian subcontinent during the 19th century. Despite their adaptability to a variety of habitats, their populations have declined due to deforestation and human settlements. By the late 1970s no more than 42,000 remained; in the early 1980s this figure fell to below 30,000. The species is still threatened by deforestation and ivory hunters. Listed in CITES Appendix I.

Order
PERISSODACTYLA (Odd-toed ungulates)

African Wild Ass, *Equus africanus* NORTHEAST AFRICA Threat: warfare, hunting (food), and drought.

Grevy Zebra, *Equus grevyi* ETHIOPIA, KENYA (possibly SOMALIA) Threat: hunting (skin) and warfare.

Indian Wild Ass, *Equus hemionus khur* PAKISTAN, INDIA Threat: uncertain, possibly disease, habitat loss (land development), drought, and human disturbance.

Przewalski Horse, *Equus przewalskii* CHINA, MONGOLIA Threat: hunting. Possibly extinct in the wild.

Cape Mountain Zebra, *Equus zebra zebra* SOUTH AFRICA Threat: habitat loss and competition with domestic stock (leading to persecution).

Malayan Tapir, *Tapirus indicus* SOUTHEAST ASIA Threat: deforestation (commercial logging, oil exploration, rubber and rice plantations, mining, settlement).

Northern Square-lipped Rhinoceros, *Ceratotherium simum cottoni* NORTHWEST ZAIRE, SOUTHERN SUDAN Also known as the Northern White Rhinoceros, this subspecies is severely threatened by poaching for its valuable horn and by the devastation caused by warfare. Trade in the horn amounts to around 1.4 tons per year—half of the horn total leaving Africa. In 1984 its population was some 500 in the wild but by 1988 this had fallen to less than 50—and these only in Zaire. Listed in CITES Appendix I.

Black Rhinoceros, *Diceros bicornis* AFRICA (south of the Sahara) Threat: poaching (horn), loss of habitat (settlements), and drought.

Sumatran Rhinoceros, *Dicerorhinus sumatrensis* SOUTHEAST ASIA Threat: overhunting and deforestation (commercial logging, settlements).

Javan Rhinoceros, *Rhinoceros sondaicus* JAVA Threat: overhunting, possibly disease.

Great Indian Rhinoceros, *Rhinocereos unicornis* INDIA, NEPAL Threat: poaching (horn) and loss of habitat (agricultural encroachment, flooding, erosion).

Order
ARTIODACTYLA (Even-toed ungulates)

Pygmy Hog, *Sus salvanius* INDIA, NEPAL (possibly BANGLADESH). Threat: loss of habitat (encroachment, settlement) and hunting.

Visayan Spotted Deer, *Cervus alfredi* VISAYAN IS. (Philippines) Threat: uncertain.

Swamp Deer, *Cervus duvauceli* INDIA, NEPAL Threat: hunting and domestic stock grazing.

Bactrian Deer, *Cervus elaphus bactrianus* AFGHANISTAN, USSR Threat: hunting, habitat loss (flooding from dams, stock grazing, cultivation), and predation (wolves).

Corsican Red Deer, *Cervus elaphus corsicanus* SARDINIA (possibly CORSICA) Threat: poaching and loss of habitat (burning, road construction, stock grazing).

Hangul, *Cervus elaphus hanglu* VALE OF KASHMIR (India) Threat: poaching and stock grazing.

Shou, *Cervus elaphus wallichi* CHINA, BHUTAN Threat: hunting.

Yarkand Deer, *Cervus elaphus yarkandensis* CHINESE TURKESTAN Threat: hunting and probable habitat loss.

Manipur Brow-antlered Deer, *Cervus eldi eldi* MANIPUR (India) Threat: hunting and human disturbance.

Thailand Brown-antlered Deer, *Cervus eldi siamensis* SOUTHEAST ASIA Threat: hunting and habitat loss (warfare, defoliation).

Shansi Sika, *Cervus nippon grassianus* CHINA Threat: hunting (antlers) and deforestation (agriculture).

Ryukyu Sika, *Cervus nippon keramae* RYUKYU IS. (Japan) Threat: lack of water and good-quality food in dry years.

South China Sika, *Cervus nippon kopschi* CHINA Threat: hunting (antlers).

North China Sika, *Cervus nippon mandarinus* CHINA Threat: hunting and deforestation.

Formosan Sika, *Cervus nippon taiouanus* TAIWAN Threat: hunting and habitat loss (agriculture).

Persian Fallow Deer, *Dama mesopotamica* MESOPOTAMIA Threat: hunting and loss of habitat (woodcutting, stock grazing, agriculture, modernization).

South Andean Huemul, *Hippocamelus bisulcus* SOUTHERN ANDES (South America) Threat: hunting, disease, competition with domestic stock and introduced deer, and habitat loss.

Fea Muntjac, *Muntiacus feae* BURMA, THAILAND Threat: hunting (food).

Argentinian Pampas Deer, *Ozotoceros bezoarticus celer* ARGENTINA Threat: hunting, disease, loss of habitat, competition with cattle and horses for forage and water, and hunting.

Wild Yak, *Bos grunniens* CENTRAL ASIA Threat: uncontrolled hunting.

Kouprey, *Bos sauveli* INDOCHINA Threat: hunting (food, horns) and loss of habitat (warfare).

Wild Asiatic Water Buffalo, *Bubalus bubalis* INDIA, NEPAL. Threat: habitat loss, disease, poaching, and competition with cattle.

Lowland Anoa, *Bubalus depressicornis* SULAWESI (Indonesia) Threat: deforestation and uncontrolled hunting.

Mountain Anoa, *Bubalus quarlesi* SULAWESI (Indonesia) Threat: deforestation and uncontrolled hunting.

Tamaraw, *Bubalus mindorensis* MINDORO IS. (Philippines) Threat: deforestation and uncontrolled hunting.

Western Giant Eland, *Taurotragus derbianus derbianus* WEST AFRICA Threat: uncontrolled hunting, disease, and probable loss of habitat.

Addax, *Addax nasomaculatus* SAHARA/SAHEL In 1982 population figures were below 2,000 and falling, leading to the belief that the addax was dangerously close to extinction. The few that survive live in desert regions far from water sources. Its decline has been caused by overhunting and loss of its preferred scrub-grass habitat due to overgrazing of domestic livestock belonging to nomadic tribes. Listed in CITES Appendix I.

Black-faced Impala, *Aepyceros melampus petersi* ANGOLA, NAMIBIA Threat: hunting and human disturbance.

Swayne Hartebeest, *Alcelaphus buselaphus swaynei* ETHIOPIA, SOMALIA Threat: hunting and loss of habitat (overgrazing of domestic livestock, settlement, agriculture).

Tora Hartebeest, *Alcelaphus buselaphus tora* EGYPT, ETHIOPIA, SUDAN Threat: hunting, habitat degradation, and disease.

Jentink Duiker, *Cephalophus jentinki* IVORY COAST, LIBERIA Threat: hunting and deforestation.

Cuvier Gazelle, *Gazella cuvieri* NORTHWEST AFRICA Threat: hunting and habitat degradation (overgrazing by domestic livestock, forest plantations).

Saudi Goitred Gazelle, *Gazella subguttorosa marica* ARABIA Threat: hunting and habitat degradation (overgrazing by domestic livestock).

Giant Sable Antelope, *Hippotragus niger variani* ANGOLA Threat: habitat loss (warfare, agriculture, pasturelands, burning).

Zanzibar Suni, *Neotragus moschatus moschatus* ZANZIBAR Threat: hunting and habitat loss.

Scimitar-horned Oryx, *Oryx dammah* SAHARA/SAHEL Threat: desertification of habitat (overgrazing by domestic livestock) and hunting.

Arabian Oryx, *Oryx leucoryx* ARABIA, MIDDLE EAST Threat: hunting (food, hide, medicines) and human disturbance (related to the oil industry).

Mountain Nyala, *Tragelaphus buxtoni* ETHIOPIA Threat: habitat destruction (woodland burning, agriculture), drought, hunting, and competition with domestic livestock.

Sumatran Serow, *Capricornis sumatraensis sumatraensis* SUMATRA Threat: hunting (food, hide, horn) and habitat loss.

Arabian Tahr, *Hemitragus jayakari* OMAN, UNITED ARAB EMIRATES Threat: hunting and competition with domestic goats for food.

Straight-horned Markhor, *Capra falconeri megaceros* AFGHANISTAN, PAKISTAN Threat: habitat degradation (overgrazing by domestic sheep).

Pyrenean Ibex, *Capra pyrenaica pyrenaica* SPAIN Threat: hunting.

Walia Ibex, *Capra walie* ETHIOPIA Threat: loss of habitat (agriculture, overgrazing by domestic livestock).

Chartreuse Chamois, *Rupicapra rupicapra cartusiana* FRANCE Threat: poaching, loss of grazing habitat (sheep), disturbance from tourists, competition for food and space with introduced herbivores (roe deer, red deer, mouflon), and hybridization with R.r. rupicapra (the common chamois).

Endangered species: birds

Order
PODICIPEDIFORMES (Grebes)

Alaotra Grebe, *Tachybaptus rufolavatus* MADAGASCAR Threat: introduced fish (tilapia, bass), which prey on chicks, and hybridization with other grebes.

Junin Grebe, *Podiceps trazanowskii* PERU Threat: Pollution from copper mine washings and changes in water level.

Order
PROCELLARIIFORMES (Albatrosses and petrels)

Amsterdam Albatross, *Diomedea amsterdamensis* AMSTERDAM IS. (Indian Ocean) Threat: habitat loss (fires, overgrazing by cattle) and predation by introduced mammals (cats, rats).

Short-tailed Albatross, *Diomedea albatrus* JAPAN Threat: low breeding potential.

Mascarene Black Petrel, *Pterodroma aterrima* MAURITIUS Threat: predation by introduced mammals (cats, dogs, rats), hunting (food), and possible pesticide pollution of the sea.

Cahow, *Pterodroma cahow* BERMUDA Threat: predation by rats, disturbance from nearby military base, competition for nest sites with tropicbirds, and tarring of plumage from oil slicks.

Magenta Petrel, *Pterodroma magentae* CHATHAM IS. (New Zealand) Threat: predation by introduced mammals (cats, rats) and forest deterioration by introduced herbivores (brush-tailed opossums and feral cattle, sheep, pigs).

Dark-rumped Petrel, *Pterodroma phaeopygia* GALAPAGOS IS., HAWAII Threat: predation by introduced mammals (black rats, dogs, pigs) and short-eared owls, and damage to nest burrows by trampling cattle.

Black Petrel, *Procellaria parkinsoni* NEW ZEALAND Threat: predation by introduced cats.

Order
CICONIIFORMES (Storks and herons)

Oriental White Stork, *Ciconia boyciana* CHINA, JAPAN, SOUTH KOREA, USSR Threat: mercury pesticides, drainage, and agriculture.

Northern Bald Ibis, *Geronticus eremita* NORTHWEST AFRICA, TURKEY, ETHIOPIA, NORTH YEMEN Threat: hunting, nest site disturbance, pesticides, and development of land for agriculture.

Crested Ibis, *Nipponia nippon* CHINA, JAPAN Threat: hunting and deforestation.

Order
FALCONIFORMES (Birds of prey)

California Condor, *Gymnogyps californianus* USA Threat: uncertain, possibly food shortage, pesticides, and human disturbance.

Spanish Imperial Eagle, *Aquila aldalberti* SPAIN, PORTUGAL The eagle is restricted to central, western, and southern Spain, and much of Portugal. Fewer than 100 breeding pairs remain—four-fifths of these in Spain, particularly in the Coto Donana National Park, south of Seville. The species is threatened by habitat loss due to forest clearance and overgrazing by livestock, poisoning of mammalian predators of game species, shooting, pesticides, and flying into overhead power cables. Listed in CITES Appendix I.

Madagascar Fish Eagle, *Haliaeetus vociferoides* MADAGASCAR Threat: persecution.

Madagascar Serpent Eagle, *Eutriorchis astur* MADAGASCAR Threat: forest destruction.

Philippine Eagle, *Pithecophaga jefferyi* PHILIPPINES Threat: forest clearance, hunting, and capture (export trade).

Mauritius Kestrel, *Falco punctatus* MAURITIUS Threat: forest destruction and predation by introduced mammals (macaque monkeys and feral cats).

Order
GALLIFORMES (Pheasants and Curassows)

White-winged Guan, *Penelope albipennis* PERU Threat: habitat loss (charcoal burning) and hunting (food).

Cauca Guan, *Penelope persipicax* COLOMBIA Threat: forest destruction.

Black-fronted Piping Guan, *Pipile jacutinga* ARGENTINA, BRAZIL, PARAGUAY Threat: forest destruction (logging) and hunting.

Horned Guan, *Oreophasis derbianus* GUATEMALA, MEXICO Threat: habitat loss (agriculture, pastureland) and hunting.

Alagoas Curassow, *Mitu mitu* BRAZIL Threat: forest destruction and hunting.

Red-billed Curassow, *Crax blumenbachii* BRAZIL Threat: forest destruction and hunting.

Gorgeted Wood Quail, *Odontophorus strophium* COLOMBIA Threat: uncertain.

Djibouti Francolin, *Francolinus ochropectus* DJIBOUTI Threat: forest destruction.

Western Tragopan, *Tragopan melanocephalus* INDIA, PAKISTAN Threat: hunting, trapping, forest destruction, and disturbance (humans, goats).

Cabot's Tragopan, *Tragopan caboti* CHINA Threat: forest destruction (agriculture) and persecution.

Chinese Monal, *Lophophorus lhuysii* CHINA Threat: overhunting.

Brown Eared-Pheasant, *Crossoptilon mantchuricum* CHINA Threat: persecution and deforestation.

Cheer Pheasant, *Catreus wallichi* INDIA, NEPAL, PAKISTAN Threat: habitat loss and hunting.

Elliot's Pheasant, *Syrmaticus ellioti* CHINA Threat: forest destruction and hunting.

White-breasted Guineafowl, *Agelastes meleagrides* WEST AFRICA Threat: hunting and forest destruction (logging).

Order
GRUIFORMES (Cranes, rails, and bustards)

Whooping Crane, *Grus americana* CANADA, USA Threat: destruction, disturbance, and pollution of wetland habitat.

Lord Howe Island Woodhen, *Tricholimnas sylvestris* LORD HOWE IS. (Australia) Threat: habitat loss (overgrazing by feral goats, pigs) and predation of eggs by rats.

Barred-wing Rail, *Nesoclopeus peociloptera* FIJI IS. Threat: predation by introduced mongooses.

Takahe, *Notornis mantelli* NEW ZEALAND Threat: competition for food with introduced deer and predation by introduced stoats.

Kagu, *Rhynochetos jubatus* NEW CALEDONIA Threat: predation by introduced mammals (dogs, cats, pigs, rats), habitat loss (nickel mining), and trapping.

Bengal Florican, *Houbaropsis bengalensis* INDIA, KAMPUCHEA, NEPAL Threat: habitat loss (agriculture) and human disturbance.

Lesser Florican, *Sympheotides indica* INDIA Threat: habitat loss (agriculture).

Order
CHARADRIIFORMES (Plovers, gulls, auks)

Chatham Island Oystercatcher, Haematopus chathamensis CHATHAM IS. (New Zealand) Threat: habitat loss.

Black Stilt, *Himantopus novaezeelandia* NEW ZEALAND Threat: habitat loss (dams, irrigation).

New Zealand Shore Plover, *Thinornis novaeseelandia* NEW ZEALAND Threat: predation by mammals.

Eskimo Curlew, *Numenius borealis* CANADA, USA Threat: loss of feeding habitat (agriculture) and shooting of migrants.

Order
COLUMBIFORMES (Pigeons)

Pink Pigeon, *Nesoenas mayeri* MAURITIUS The pigeon's population has been critically low since 1960 and in 1976 there were only 30 birds left. Some 100 zoo-bred individuals have since been introduced to the wild. Main threats include forest destruction, predation by introduced animals (black rats, cats, mongooses, crab-eating macaques, Indian mynahs) and native birds, hunting, late winter food shortages, and cyclones. Listed in CITES Appendix III for Mauritius.

Marquesas Pigeon, *Ducula galeata* MARQUESAS IS. (South Pacific) Threat: hunting (food) and habitat loss (overgrazing by livestock).

Order
PSITTACIFORMES (Parrots)

Paradise Parrot, *Psephotus pulcherrimus* AUSTRALIA Threat: uncertain, probably habitat loss and capture (pet trade).

Ground Parrot, *Pezoporus wallicus* AUSTRALIA Threat: habitat loss (settlements).

Mauritius Parakeet, *Psittacula eques* MAURITIUS Threat: deforestation, competition for nest sites with introduced birds (mynahs), and nest predation by introduced mammals (macaque monkeys, black rats).

Maroon-fronted Parrot, *Rhynchopsitta terrisi* MEXICO Threat: habitat loss (logging), hunting (food), and capture (pet trade).

Puerto Rican Amazon, *Amazona vittata* PUERTO RICO Threat: habitat loss, heavy rainfall, disease, competition from other birds, and predation (red-tailed hawks, rats).

Red-tailed Amazon, *Amazona brasiliensis* BRAZIL Threat: deforestation and capture (pet trade).

Red-necked Amazon, *Amazona arausiaca* DOMINICA (Caribbean) Threat: hunting and competition for nest sites with other birds.

St Vincent Amazon, *Amazona guildingii* ST VINCENT (Caribbean) Threat: hunting and capture (pet trade).

Imperial Amazon, *Amazona imperialis* DOMINICA (Caribbean) Threat: hunting (food, sport) and habitat loss.

Kakapo, *Strigops habroptilus* NEW ZEALAND Threat: predation by introduced mammals (stoats, rats) and competition for food with other herbivores (deer, brush-tailed opossums, chamois).

Order
CORACIIFORMES (Woodpeckers, toucan)

Helmeted Woodpecker, *Dryocopus galeatus* BRAZIL, ARGENTINA, PARAGUAY Threat: forest destruction (agriculture, livestock ranching).

Ivory-billed Woodpecker, *Campephilus principalis* USA, CUBA Most of the mature bottomland, hardwood swamp forest that this woodpecker inhabits has been cleared and only isolated stands remain in its ill-defined range in southeast USA. Its population is uncertain and sightings or tape recordings of its call are little publicized or else treated with scepticism. Threats include continued forest clearance, commercial collection of birds and eggs, and other human disturbances.

Imperial Woodpecker, *Campephilus imperialis* MEXICO Threat: deforestation (logging) and hunting.

Okinawa Woodpecker, *Sapheopipo noguchii* JAPAN Threat: habitat loss (fires, forestry).

Order
PASSERIFORMES (Perching birds)

Black-hooded Antwren, *Myrmotherula erythronotos* BRAZIL Threat: forest destruction.

Fringe-backed Fire-eye, *Pyriglena atra* BRAZIL Threat: forest destruction (agriculture, industry, settlements).

Gurney's Pitta, *Pitta gurneyi* BURMA, THAILAND Threat: forest destruction.

New Zealand Bush Wren, *Xenicus longpipes* NEW ZEALAND Threat: predation by introduced mammals (rats, cats).

Raso Lark, *Alauda razae* RASO (Cape Verde Is.) Threat: predation by introduced mammals (rats) and prolonged drought.

White-breasted Thrasher, *Ramphocinclus brachyurus* MARTINIQUE, ST LUCIA (Caribbean) Threat: predation by introduced mammals (mongooses, rats).

Thyolo Alethe, *Alethe choloensis* MOZAMBIQUE, MALAWI Threat: forest clearance and disturbance.

Seychelles Magpie-robin, *Copsychus sechellarum* SEYCHELLES Threat: predation by introduced mammals (cats) and competition with/nest predation by introduced birds.

Puaiohi, *Myadestes palmeri* HAWAII Threat: habitat disturbance (encroachment by exotic plants, introduced herbivores), predation by introduced mammals and avian diseases.

Taita Thrush, *Turdus helleri* KENYA Threat: habitat loss (tree felling, agriculture, plantations, wood for fuel).

Rodrigues Warbler, *Acrocephalus rodericanus* RODRIGUES IS. (Mauritius) Threat: habitat loss, disturbance by cyclones, and predation by black rats.

Aldabra Warbler, *Nesillas aldabranus* ALDABRA (Seychelles) Threat: habitat disturbance (goats, tortoises) and predation by introduced rats.

Long-legged Warbler, *Trichocichla rufa* FIJI IS. Threat: predation by introduced mammals.

Chatham Island Black Robin, *Petroica traversi* CHATHAM IS. (New Zealand) Threat: human disturbance.

Banded Wattle-eye, *Platysteira laticincta* CAMEROON Threat: habitat loss (forest clearance, cultivation, woodcutting, fires, and overgrazing by cattle, goats, sheep, horses).

Marungu Sunbird, *Nectarina prigoginei* ZAIRE Threat: habitat loss (logging, erosion of stream banks due to overgrazing by cattle).

White-breasted White-eye, *Zosterops albogularis* NORFOLK IS. (Australia) Threat: deforestation, shooting, and predation by introduced mammals (rats).

Kaual Oo, *Moho braccatus* HAWAII Threat: habitat loss (introduced herbivores, encroachment by exotic plants) and introduced predators (black rats).

Bachman's Warbler, *Vermivora bachmanii* CUBA, USA Threat: habitat loss (logging, agriculture).

Kirtland's Warbler, *Dendroica kirtlandii* BAHAMAS, USA Threat: habitat loss and brood parasites (brown-headed cowbirds, now removed).

Semper's Warbler, *Leucopeza semperi* ST LUCIA (Caribbean) Threat: uncertain, probably predation by introduced mongooses.

Akialoa, *Meignathus obscurus* HAWAII Threat: habitat loss (overgrazing by cattle, goats, pigs), introduced diseases, and introduced rats.

Nukupuu, *Hemignathus lucidus* HAWAII Threat: habitat disturbance (encroachment by exotic plants and animals).

Akiapolaau, *Hemignathus munroi* HAWAII Threat: habitat loss and introduced diseases, predators, and birds.

Ou, *Psittirostra psittacea* and **Palila** *Psittirostra bailleui* HAWAII Threat: habitat loss and introduced predators, bird competitors, and diseases.

Clarke's Weaver, *Ploceus golandi* KENYA Threat: forest destruction (agriculture, exploitation).

Rodrigues Fody, *Foudia flavicans* RODRIGUES IS. (Mauritius) Threat: deforestation and introduced bird competitors.

Bali Starling, *Leucospar rothschildi* INDONESIA Threat: capture (pet trade), habitat loss (settlements), and competition with other starlings.

Hawiian Crow, *Corvus tropicus* HAWAII Threat: habitat loss (overgrazing by feral pigs, cattle, goats), introduced diseases, and predation by introduced rats.

Endangered species: fishes

Order
ACIPENSERIFORMES (Sturgeons)

Common Sturgeon, *Acipenser sturio* EUROPE Threat: uncertain.

Pallid Sturgeon, *Scaphirhynchus albus* USA An inhabitant of the waters of the Mississippi and Missouri valleys, its numbers have declined drastically since 1900. Several states in its range have provided protection. Main threats involve modifications of its habitat: dams and water canalization changes, siltation, and pollution of spawning and feeding grounds. Other threats include hybridization with *S. platyrhynchus* and incidental harvest of young individuals in commercial sturgeon catches.

Order
CYPRINIFORMES (Carps)

Berg River Redfin, *Barbus burgi* SOUTH AFRICA Threat: water pollution, habitat changes (irrigation, drainage, mining/quarrying, canalization), competition, and predation.

Barbus srilankensis SRI LANKA Threat: habitat changes (siltation, mining/quarrying).

Flatjaw Minnow, *Dionda mandibularis* MEXICO Threat: competition from four introduced fish species (e.g. *Gambusia panuco*, *Poecilia mexicana*) and water table changes (irrigation).

Bonytall, *Gila elegans* USA Threat: salt pollution, habitat changes (dams, irrigation, drainage, fall in water temperature), hybridization, predation, and competition with exotic species.

Green Labeo, *Labeo fisheri* SRI LANKA Threat: habitat changes (dams), overfishing, and explosives as means of capture.

White River Spinedace, *Lepidomeda albivallis* USA Threat: water pollution, habitat changes (dams, irrigation, drainage, canalization), predation, and competition with exotic species.

Cuatro Cienagas Shiner, *Notropis xanthicara* MEXICO Threat: habitat changes (drainage, canalization), tourism, predation, and competition with exotic species.

Drakensburg Minnow, *Oreodaimon quathlambae* LESOTHO Threat: habitat changes (dams, roads, soil erosion, siltation), predation, and competition with exotics.

Woodfin, *Plagopterus argentissimus* USA Threat: habitat changes (arroyo cutting, siltation, water removal, dams, soil erosion).

Tokyo Bitterling, *Tanakia tanago* JAPAN Threat: habitat loss (urbanization).

Tylognathus klatti TURKEY Threat: water table changes (irrigation).

Spotted Loach, *Lepidocephalus jonklassi* SRI LANKA Threat: habitat changes (deforestation).

Ayumodoki, *Leptobotia curta* JAPAN Threat: habitat changes (riverbank alteration).

Modoc Sucker, *Catastomus microps* USA Threat: water pollution, habitat changes (soil erosion, siltation, riverbank collapse, canalization), predation, and hybridization with exotic species.

Shortnose Sucker, *Chasmistes brevirostris* USA Threat: habitat changes (agriculture, dams, drainage), predation, competition, and hybridization with exotic species.

Cui-ui, *Chasmistes cujus* USA Threat: water pollution, disease, overfishing, and habitat changes (agriculture, logging, dams, siltation, irrigation, fall in water temperature).

June Sucker, *Chasmistes liorus* USA Threat: exploitation (food, sport), drought, water table changes (irrigation), predation, competition, and hybridization with exotic species.

Order
SILURIFORMES (Catfishes)

Tollo de Agua, *Diplomystes chilensis* CHILE Threat: competition with exotic species and naturally low populations.

Quachita Madtom, *Noturus lachneri* USA Threat: water pollution and habitat changes (gravel extraction, logging, urbanization, dams, canalization).

Barnard's Rock Catfish, *Austroglanis barnardi* SOUTH AFRICA Threat: habitat changes (drainage, siltation, canalization), predation by exotic species, and naturally low populations.

Nekogigi, *Coreobagrus ichikawai* JAPAN Threat: habitat changes (dams).

Order
SALMONIFORMES (Salmon)

Swan Galaxias, *Galaxias fontanus* TASMANIA Threat: predation by exotic species.

Atlantic Whitefish, *Coregonus candensis* GREAT LAKES (North America) Threat: acid rain, overfishing, habitat changes (dams), predation, and competition with exotic species.

Gila Trout, *Salmo gilae* USA Formerly ubiquitous and abundant, the trout is now restricted to five headwater streams of two New Mexico rivers, the San Francisco and the Gila. Its estimated total population is 10,000. Main threats include hybridization with introduced rainbow trout (*S. gairdneri*), habitat changes due to agriculture and forestry, competition with introduced brown and brook trout, as well as droughts, floods, and fires. Listed as endangered by the U.S. Fish and Wildlife Service.

Order
CYPRINODONTIFORMES

Ginger Pearlfish, *Cynolebias marmoratus*, **Splendid Pearlfish** *C. splendens*, **Opalescent Pearlfish** *C. opalescens* BRAZIL Threat: habitat loss (land reclamation, agriculture).

Valencia Toothcarp, *Valencia hispanica* SPAIN Threat: habitat loss.

Pescos Gambusia, *Gambusia nobilis* USA Threat: habitat changes (dams, irrigation, drainage), predation, competition, and hybridization with exotic and natural species.

Monterrey Platyfish, *Xiphophorus couchianus* MEXICO Threat: habitat changes (drainage, flooding, urbanization) and hybridization with exotic species.

Devil's Hole Pupfish, *Cyprinodon diabolis* USA Threat: habitat changes (agriculture, drainage, siltation), human disturbance (vandals, divers), and naturally low populations.

Desert Pupfish, *Cyprinodon macularius* MEXICO, USA Threat: water pollution, habitat changes (dams, irrigation, dredging, urbanization, agriculture), disease, competition, and hybridization with exotic species.

Order
PERCIFORMES (Perchlike fishes)

Watercress Darter, *Etheostoma nuchale* USA Threat: water pollution, habitat changes (removal of vegetation, drainage, industrial practices, roads), and disease.

Conasauga Logperch, *Percina jenkinsi* USA Threat: water pollution and habitat changes (logging, agriculture, dams, siltation, industrial practices, canalization, urbanization).

Otjikoto Tilapia, *Tilapia guinasana* NAMIBIA Threat: water pollution and habitat changes (irrigation, drainage, mining/quarrying, agriculture).

Endangered species: reptiles and amphibians

REPTILES

Order
TESTUDINES (Turtles and tortoises)

Western Swamp Turtle, *Pseudemydura umbrina* WESTERN AUSTRALIA Threat: habitat loss (urban development, agriculture, wildfires), a drier climate, and predation by introduced red foxes.

Green Turtle, *Chelonia mydas* TROPICAL SEAS Threat: exploitation of adults and eggs (food), juveniles (stuffed as curios), and adults (hide, oil); also shrimp trawl nets.

Hawksbill Turtle, *Eretmochelys imbricata* TROPICAL SEAS Threat: exploitation (stuffed as curios, tortoiseshell, eggs and adults for food).

Kemp's Ridley, *Lepidochelys kempii* GULF OF MEXICO Threat: exploitation of eggs, juveniles, and adults (food), predation of eggs by coyotes, shrimp trawl nets, and pollution.

Olive Ridley, *Lepidochelys olivacea* TROPICAL SEAS Threat: exploitation of adults (food, skin), massive harvest of eggs (food), and shrimp trawl nets.

Leatherback, *Dermochelys coriacea* TROPICAL and TEMPERATE SEAS The world population of breeding females was estimated in 1981 to be some 104,000. The main threat to the species is the excessive harvest of its eggs for food. Lesser threats include the taking of adults for food, oil, shark bait, and medicinal purposes. The leatherback is increasingly caught in shrimp trawl nets and squid drift nets. Its nesting habitats are threatened by tourism and marine erosion. Listed in CITES Appendix I.

River Terrapin, *Batagur baska* SOUTHEAST ASIA Threat: exploitation of adults and eggs (food), habitat destruction (deforestation, mining, sand removal), and human disturbance.

South American River Turtle, *Podocnemis expansa* AMAZON BASIN Threat: exploitation of eggs (food) and adults (food, oil), natural flooding of nests, and habitat loss (forest clearance, dams).

Bolson Tortoise, *Gopherus flavomarginatus* MEXICO Threat: exploitation (food), habitat destruction (overgrazing, agriculture, irrigation), and capture (export trade).

Order
CROCODYLIA (Crocodiles and alligators)

Chinese Alligator, *Alligator sinensis* CHINA Threat: habitat loss (settlements), drought, exploitation (hide), and natural flooding.

Broad-nosed Caiman, *Caiman latirostris* SOUTHERN SOUTH AMERICA Threat: poaching (hide) and habitat loss (drainage, forest clearance, dams, agriculture, ranching, logging, pollution).

Black Caiman, *Melanosuchus niger* SOUTH AMERICA Threat: poaching (hide), persecution, and habitat loss (logging, agriculture, cattle ranching).

American Crocodile, *Crocodylus acutus* CARIBBEAN, CENTRAL AMERICA, USA Threat: poaching (hide) and habitat loss (salinity changes in water, cultivation, drainage).

Orinoco Crocodile, *Crocodylus intermedius* COLOMBIA, VENEZUELA Threat: exploitation (hide) and persecution.

Philippines Crocodile, *Crocodylus mindorensis* PHILIPPINES Threat: exploitation (hide) and habitat loss (agriculture/aquaculture operations).

Estuarine Crocodile, *Crocodylus porosus* ASIA, AUSTRALIA, WESTERN PACIFIC The population of this large crocodile is uncertain. It is mainly threatened by excess commercial hunting for its hide, which produces the finest leather. Other threats include mangrove forest destruction, collection of eggs and young for "farm" rearing and food, and persecution because of occasional human fatalities. Listed in CITES Appendix I.

Slamese Crocodile, *Crocodylus siamensis* SOUTHEAST ASIA Threat: exploitation (hide) and habitat loss (rice paddies).

False Gavial, *Tomistoma schlegelii* SOUTHEAST ASIA Threat: exploitation (hide), collection of young "farm" rearing), and habitat loss (timber, rice paddies, settlement, pollution).

Gavial, *Gavialis gangeticus* SOUTHEAST ASIA Threat: habitat loss (dams, irrigation), exploitation (hide), collection of eggs (food), and drowning in nylon set-nets.

Order
SAURIA (Lizards)

Rodrigues Day Gecko, *Phelsuma edwardnewtonii* RODRIGUES IS. (Mauritius) Threat: predation by introduced mammals (cats, rats).

Culebra Island Giant Anole, *Anolis roosevelti* CULEBRA IS. (Puerto Rico) Threat: habitat loss (land development).

San Joaquin Leopard Lizard, *Gambelia silus* SOUTHERN CALIFORNIA (USA) Threat: habitat loss (agriculture, water control operations).

St Croix Ground Lizard, *Ameiva polops* US VIRGIN IS. Threat: habitat loss and predation by introduced monogooses.

Order
SERPENTES (Snakes)

Round Island Boa, *Bolyeria multocarinata* ROUND IS. (Mauritius) Threat: habitat loss (overgrazing by rabbits and goats leading to soil erosion).

Round Island Keel-scaled Boa. *Casarea dussumieri* ROUND IS. (Mauritius) Threat: habitat loss (overgrazing by rabbits and goats leading to soil erosion).

Puerto Rican Boa, *Epicrates inornatus* PUERTO RICO Threat: predation by introduced mongooses, persecution, auto accidents, and oil spills.

San Francisco Garter Snake, *Thamnophis sirtalis tetrataenia* WESTERN USA Threat: habitat loss (land development, drainage systems).

Latifi's Viper, *Vipera latifii* IRAN Threat: habitat loss (hydroelectric power operations).

AMPHIBIANS

Order
ANURA (Frogs and toads)

Houston Toad, *Bufo houstonensis* TEXAS (USA) Threat: habitat loss (agriculture, urbanization, drainage) and hybridization with other toad species.

Conondale Gastric-brooding Frog, *Rheobatrachus silus* QUEENSLAND (Australia) Threat: habitat loss (land clearance) and disturbance.

Vegas Valley Leopard Frog, *Rama fisheri* USA Threat: habitat loss (capping of springs) and competition with introduced amphibians.

Order
CAUDATA (Salamander)

Desert Slender Salamander, *Batrachoseps aridus* CALIFORNIA (USA) Threat: habitat loss (destruction of canyon limestone sheeting).

Texas Blind Salamander, *Typhlomolge rathbuni* TEXAS (USA) Threat: overcollection, capping of wells, drainage, water pollution, and a reduction of its aquatic invertebrate diet.

Endangered species: invertebrates

Phylum Platyhelminthes *Class* **Turbellaria** *Order*
TRICLADIDA

Holsinger's Groundwater Planarian, *Sphalloplano holsingeri*, and **Bigger's Groundwater Planarian**, *Sphalloplano subtilis* VIRGINIA (USA) (probably extinct) Threat: housing development.

Phylum Mollusca *Class* **Gastropoda** *Order*
MESOGASTROPODA

Point of Rocks Spring Snail, *Fluminicola erythopoma* USA Threat: habitat loss (drainage, capping of springs, roads).

Giant Columbia River Spire Snail, *Lithoglyphus columbiana* USA Threat: water pollution and dam construction.

Spiny River Snail, *Io fluvialis* USA Threat: water pollution and dam construction.

Order
BASOMMATOPHORA

Tasmanian Freshwater Limpet, *Ancylastrum cumingianus* TASMANIA Threat: introduced trout as a sport fish.

Order
STYLOMMATOPHORA

Little Agate Shells, *Achatinella* [19 spp.] HAWAII Threat: habitat destruction (forest fires), introduced plant and animal species (the carnivorous snail, *Euglandina rosea*), and collection.

Partulina confusa HAWAII Threat: deforestation, collection (trophies, specimens), and predation.

Viviparous Tree Snails, *Samoana diaphana, S. solitaria, Partula hebe* SOCIETY IS. (South Pacific) Threat: the carnivorous snail, *Euglandina rosea*, which was introduced to Moorea Island to combat the previously introduced Giant African Snail, *Achatina fulica*. *E. rosea* has already caused the extinction of nine *Partula* species from Moorea Island.

Granulated Tasmanian Snail, *Anoglypta launcestonensis* TASMANIA Threat: land clearances (raising of pasture grasses and monocultures of pine, wheat, and other crops).

Thaumatodon hystricelloides WESTERN SAMOA (South Pacific) Threat: predation by ants.

Virginia Fringed Mountain Snail, *Polygyriscus virginianus* USA Threat: herbicides, road building, and mining/quarrying.

Flat-spired Three-toothed Snail, Triodopsis

platysayoides USA Threat: human disturbance (recreation).

Class **Bivalva**
Order **UNIONOIDA**

Birdwing Pearly Mussel Birdwing, *Conradilla caelata* USA Threat: water pollution and dam construction.

Dromedary Pearly Mussel, *Dromus dromas* USA Threat: water pollution.

Epioblasma Mussels, *Epioblasma* [15 spp.] USA. Most threatened by water pollution, large-scale waterway systems, and commercial exploitation. The **Green-blossom Pearly Mussel**, *E. torulosa gubernaculum*, the **Tan-blossom Pearly Mussel**, *E.t. rangiana*, and the **Tubercled-blossom Pearly Mussel**, *E.t. torulosa*, are particularly threatened by water impoundments, domestic sewage treatment plant effluents, industrial outfalls, agricultural silt and pesticide run-off, dredging, and canalization of streams. **The Tan Riffle Shell Mussel**, *E. walkeri*, is particularly threatened by mine acids, municipal wastes, lead and mercury pollution, dams, and stream canalization.

Fine-rayed Pigtoe Pearly Mussel, *Fusconaia cuneolus*, **Shiny Pigtoe Pearly Mussel**, *F. edgariana* USA Threat: water pollution.

Cracking Pearly Mussel, *Hemistena lata* USA Threat: water pollution.

Higgin's Eye Pearly Mussel, *Lampsilis higginsi* USA Threat: agriculture.

Subphylum **Crustacea**
Order
ISOPODA (Isopods)

Socorro Isopod, *Thermosphaeroma thermophilum* NEW MEXICO (USA) A population of about 2,500 lives in a single thermal outflow from Sedillo Spring near Socorro City. The isopod is of scientific interest because of its evolutionary adaptation and behavior. It is threatened by municipal and private water developments (which could divert water away from its habitat), and water pollution. Listed as endangered by the U.S. Fish and Wildlife Service.

Order
DECAPODA

California Freshwater Shrimp, *Syncaris pacifica* USA Threat: competition with introduced species.

Nashville Crayfish, *Orconectes shoupi* TENNESSEE (USA) A small but uncounted population, living in Mill Creek near Nashville, is threatened by increased siltation, water pollution, and habitat loss.

Subphylum **Uniramia** *Class* **Insecta** *Order*
ODONATA (Dragonflies)

Ecchloroletes nylepytha, E. peringueyi SOUTH AFRICA Threat: agriculture and ecological changes.

Small Damselfly, *Hemiphlebia mirabilis* VICTORIA (Australia) This primitive damselfly is the only species in its family and is possibly extinct. It relies on seasonal inundations of its habita—reedy lagoons on flood plains, Damming, drainage schemes, and agriculture have prevented flooding and pose the main threat to the species.

Amanipodagrion gilliesi TANZANIA Threat: agriculture.

Nososticta pilbara AUSTRALIA Threat: agriculture.

Platycnemis mauricana MAURITIUS Threat: drainage.

Phylum Annelida *Class* Oligochaeta *Order*
HAPLOTAXIDA

Washington Giant Earthworm, *Megascolides americanus*, **Oregon Giant Earthworm**, *Megascolides macelfreshi* USA Threat: altered food supply due to conversion of their habitat for crops and pastureland; also industrialization and urbanization.

Phylum Arthropoda *Subphylum* **Chelicerata** *Class* **Arachnida** *Order*
ARANEAE (Spiders)

No-eyed Big-eyed Spider, *Adelocosa anops* KAUAI (Hawaiian Is.) A small but uncounted population of these blind big-eyed spiders lives in caves formed by lava tubes after the volcanic eruption of Koloa. It is threatened by general habitat disturbance caused by tourist pressure, surface development, withdrawal and pollution of water, waste residue from sugarcane plantations, and deforestation (which causes loss of tree root—an important source of food to the cave fauna on which the spider feeds).

Freya's Damselfly, *Coenagrion hylas freyi* AUSTRIA, GERMANY, SWITZERLAND Threat: habitat loss (alteration of alpine lake systems leading to reduction in lakeshore *Equisetum* beds).

San Francisco Forktail Damselfly, *Ischnura gemina* SAN FRANCISCO BAY (California, USA) Threat: disturbance and modifications of habitat; also pollution.

Teinobasis alluaudi alluaudi SEYCHELLES Threat: afforestation.

Grüne Keiljungter, *Ophiogomphus cecilia* EUROPE to SIBERIA Threat: uncertain.

Florida Spiketail Dragonfly, *Cordulegaster sayi* USA Threat: urban housing development and pesticides.

Ohio Emerald Dragonfly, *Somatochlora hineana* OHIO, INDIANA (USA) Threat: uncertain. Possibly extinct.

Order
DERMAPTERA (Earwigs)

St Helena Earwig, *Labidura herculeana* ST HELENA (South Atlantic Ocean) Probably extinct. Threat: habitat loss (settlement and overgrazing by goats, rabbits), competition with three species of introduced earwigs, and predation by introduced animals (rat, giant centipede).

Order
PLECOPTERA

Otway Stonefly, *Eusthenia nothofagi* VICTORIA (Australia) Threat: habitat loss (streamside forest clearance for agriculture).

Order
ANOPLURA (Sucking lice)

Pygmy Hog Sucking Louse, *Haematopinus oliveri* NORTHWEST ASSAM (India) Threat: habitat destruction has endangered the louse and its sole host, the pygmy hog.

Order
COLEOPTERA (Beetles)

Columbia River Tiger Beetle, *Cicindela columbica* IDAHO (USA) Threat: habitat loss (dams, flooding).

American Burying Beetle, *Nicrophorus americanus* EASTERN NORTH AMERICA (possibly restricted to Rhode Island, USA, only) Threat: habitat loss (deciduous forest clearance).

Kauai Flightless Stag Beetle, *Apterochychus honoluluensis* HAWAII Threat: uncertain.

Woodruff's Dung Beetle, *Ataenius woodruffi* USA Threat: uncertain.

Hermit Beetle, *Osmoderma eremita* EUROPE Threat: loss of habitat through destruction or intensive management of ancient woodland.

Goldstreifiger, *Buprestis splendens* EUROPE Threat: loss of habitat through destruction or intensive management of ancient woodland.

Hawaiian Click Beetles, *Eopenthes* [17 spp.] HAWAII Threat: uncertain.

Cerambyx Longicorn, *Cerambys cerdo* NORTHERN and CENTRAL EUROPE Threat: destruction as a pest of oak trees and loss of habitat (woodland destruction and intensive management).

Mojave Rabbitbrush Longhorn Beetle, *Crossidius mojavensis mojavensis* USA Threat: uncertain.

Rosalia Longicorn, *Rosalia alpina* NORTHERN and CENTRAL EUROPE

Threat: loss of habitat (woodland destruction and intensive management).

Hawaiian Proterhinus Beetles, *Proterhinus* [72 spp.] HAWAII Threat: uncertain.

Hawaiian Snout Beetles, *Rhynocogonus* [22 spp.] HAWAII Threat: uncertain.

Order
DIPTERA (True flies)

Giant Torrent Midge, *Edwardsina gigantea* NEW SOUTH WALES (Australia) Threat: changes in river flow (dams) and river pollution (sewage effluent).

Tasmania Torrent Midge, *Edwardsina tasmaniensis* TASMANIA Threat: changes in river flow (hydroelectric power operations).

Belkin's Dune Tabinid Fly, *Brennania belkini* MEXICO Threat: habitat loss (encroachment by exotic plants, off-road vehicles, urban development).

Order
LEPIDOPTERA. (Butterflies and moths)

Dakota Skipper, *Hesperia dacotae* USA, CANADA Threat: habitat loss (intensive agriculture, ranching, gravel mining, settlements, irrigation).

Pawnee Montane Skipper, *Hesperia pawnee montana* USA Threat: uncertain.

Harris' Mimic Swallowtail, *Eurytides lysithous harrisianus* SOUTHEAST BRAZIL Threat: habitat loss (development of recreational area).

Queen Alexandra's Birdwing, *Ornithoptera alexandrae* PAPUA NEW GUINEA Threat: habitat loss (logging, oil palm plantations) and overcollecting.

Schaus' Swallowtail, *Papilio aristodemus ponceanus* FLORIDA KEYS (USA) Threat: habitat loss (urban development), insecticide spraying, hurricanes, droughts, frost, and overcollecting.

Papilio chikae PHILIPPINES Threat: overcollecting and habitat loss.

Taita Blue-banded Papilio, *Papilio desmondi teita* KENYA Threat: habitat loss (tree-felling, agriculture, settlements).

Homerus Swallowtail, *Papilio homerus* JAMAICA Threat: habitat loss (timber and coffee plantations) and overcollecting.

Corsican Swallowtail, *Papilio hospiton* CORSICA, SARDINIA Threat: habitat loss, commercial collecting, and destruction of food-plants (which are poisonous to sheep).

Parnassus apollo vinningensis FEDERAL REPUBLIC of GERMANY Threat: habitat loss (coniferous afforestation) and possibly overcollecting.

Golden Birdwing, *Troides aeacus kaguya* TAIWAN Threat: overcollecting.

San Bruno Elfin, *Callophrys mossii bayensis* USA Threat: uncertain.

Mission Blue, *Icaricia icarioides missionensis* USA Threat: uncertain.

Lotis Blue, *Lycaeides argyrognomon lotis* USA Threat: uncertain.

Large Copper, *Lycaena dispar* NORTHERN EUROPE Threat: loss of habitat through drainage of wetlands, flooding of valleys for reservoirs, and vegetational succession.

Dusky Large Blue, *Maculinea nausithous* EUROPE, USSR; **Scarce Large Blue**, *M. teleius* EUROPE and NORTHERN ASIA; **Scarce Large Blue**, *M.t. burdigalensis* FRANCE Threat: habitat loss (intensive agriculture discourages ants on which the large blues depend), fertilizers and herbicides, and drainage and development of wetlands.

Dickson's Copper, *Oxychaeta dicksoni* SOUTH AFRICA Threat: uncertain.

Tailless Blue, *Panchala ganesa loomisi* JAPAN Threat: uncertain.

Mission Blue, *Plebejus icarioides missionensis* USA Threat: uncertain.

Nevada Blue, *Plebicula golgus* SPAIN Threat: habitat degradation (overgrazing, ski developments).

False Ringlet, *Coenonympha oedippus* EUROPE and NORTHERN ASIA Threat: habitat loss through land drainage and grassland improvement.

Lange's Metalmark, *Apodemia mormo langei* USA Threat: uncertain.

Bay Checkerspot Butterfly, *Euphydryas editha bayensis* SAN FRANCISCO PENINSULA (California, USA) Threat: habitat loss (urban development, golf courses), drought, and pesticide spraying.

Natterer's Longwing, *Heliconius nattereri* SOUTH BRAZIL Threat: forest destruction, overcollection, and competition with other insects.

Scarce Fritillary, *Hypodryas maturna* EUROPE Threat: habitat loss through drainage of wetlands followed by afforestation.

Great Peacock Moth, *Saturnia pyri* FRANCE, SPAIN Threat: uncertain.

Prairie Sphinx Moth, *Euproserpinus wiesti* COLORADO (USA) Threat: overcollecting, insecticide spraying, oil and gas exploration, competition with other moths (white-lined sphinx, *Hyles lineata*).

Order
HYMENOPTERA (Ants, bees, wasps)

Niihau Vespid Wasp, *Odynerus niihauensis*, **Soror Vespid Wasp**, *O. soror* HAWAII Threat: uncertain.

Hawaiian Sphecid Wasp, *Deinomimesa hawaiiensis*, **Puna Sphecid Wasp**, *D. punae*; **Short-foot Sphecid Wasp**, *Ectemnius curtipes*, **Brown Cross Sphecid Wasp**, *E. fulvicrus*, **Giffard's Sphecid Wasp**, *E. giffardi*, **Haleakala Sphecid Wasp**, *E haleakalae*; **Kaual Sphecid Wasp**, *Nesomimesa kauaiensis*, *N. perkins*, **Shade-winged Sphecid Wasp**, *N. sciopteryx*, HAWAII Threat: uncertain.

INDEX

Charts

Time chart

ERA	PERIOD	EPOCH	MILLIONS OF YEARS AGO
CENOZOIC	QUATERNARY	Holocene (Recent)	0.01
		Pleistocene	2
	TERTIARY	Pliocene	5
		Miocene	25
		Oligocene	38
		Eocene	55
		Paleocene	65
MESOZOIC	CRETACEOUS		
			144
	JURASSIC		
			213
	TRIASSIC		
			248
PALEOZOIC	PERMIAN		
			286
	CARBONIFEROUS	Pennsylvanian	320
		Mississippian	360
	DEVONIAN		
			408
	SILURIAN		
			438
	ORDOVICIAN		
			505
	CAMBRIAN		
			590
PRECAMBRIAN	PROTEROZOIC EON		
			2500
	ARCHEAN EON		
			4600

In the geological time chart (above) the major units are the eras, starting with the Precambrian. Each era is further subdivided into periods and epochs. The dates on the chart indicate when each unit of time is thought to have begun.

Metric-Imperial equivalents

Length
1 centimeter = 0.393 inch
1 meter = 3.28 feet
1 kilometer = 0.621 miles

Area
1 sq centimeter = 0.155 sq inch
1 sq meter = 10.76 sq feet
1 hectare (10,000 sq m) = 2.471 acres
1 sq kilometer = 0.386 sq mile

Weight
1 gram = 0.0353 ounces
1 kilogram = 2.205 pounds
1 tonne = 1.1 tons (US)

Miscellaneous
1 kilogram/sq meter = 0.2 pounds/sq foot
°C multiplied by 9/5 plus 32 = °F
°F minus 32 multiplied by 5/9 = °C

Further reading

Anderson, S. and Jones, J. Knox Jr. *Orders and Families of Recent Mammals of the World* John Wiley, New York, 1984

Attenborough, D. *The Living Planet* Collins & BBC Publications, London, 1984

Austin, O.L. Jr. *Birds of the World* Country Life Books/Hamlyn, Twickenham, UK, 1987; Golden Press, New York, 1966

Axelrod, H.R. *African Cichlids of Lakes Malawi and Tanganyika* T.F.H. Publications, Reigate, UK, 1973

Baker, R. *Migration Paths Through Time and Space* Hodder & Stoughton, London, 1982; *The Mystery of Migration* (Ed.) Macdonald, London, 1980

Benson, L. *The Cacti of the United States and Canada* Stanford University Press, California, 1982

Blij, H.J. De *Man Shapes the Earth: A Topical Geography* Hamilton Publishing, Santa Barbara, California, 1974

Burton, J.A. *Collins Guide to the Rare Mammals of the World* Collins, London, 1987

Campbell, A.C. *The Seashore and Shallow Seas of Britain and Europe* Country Life Books/Hamlyn, Feltham, UK, 1988

Cloudsley-Thompson, J. *Animal Migration* Orbis, London, 1978

Cox, Professor B., Savage, Professor R.J.G., Gardiner, Professor B. and Dixon, D. *Macmillan Encyclopedia of Dinosaurs and Prehistoric Animals* Macmillan Publishing, New York and London, 1988

Cox, C.B. and Moore, Peter D. *Biogeography (4th Ed.)*, Blackwell Scientific Publications, Oxford, UK, 1985

Cramp, S. *Handbook of the Birds of Europe, the Middle East and North America: The Birds of the Western Palearctic (5 Vols.)* Oxford University Press, UK, 1977

Croat, T.B. *Flora of Barro Colorado Island* Stanford University Press, California, 1978

Diamond, A.W., Schreiber, R.L., Attenborough, D. and Prestt, I. *Save the Birds* Cambridge University Press, London and New York, 1987

Durrell, L. *State of the Ark* Doubleday, New York, 1986

Ehrlich, P. and A. *Extinction: The Causes and Consequences of the Disappearance of Species* Random House, New York, 1981

Erwin, D. and Picton B. *Guide to Inshore Marine Life* Immel Publishing, London, 1987

Fincham, A.A. *Basic Marine Biology* British Museum (Natural History)/Cambridge University Press, London, 1984

Fisher, J. and Peterson, R.T. *The World of Birds* Macdonald, London, 1964

Fryer, G. and Iles, T.D. *The Cichild Fishes of the Great Lakes of Africa: Their Biology and Evolution* Oliver & Boyd, Edinburgh, 1972

Gooders, J. *The World's Wildlife Paradises* David & Charles, London, 1975

Halstead, L.B. *Hunting the Past* Hamish Hamilton, London, 1982

Harrison, C. *An Atlas of the Birds of the Western Palearctic* Collins, London, 1982

Heywood, Professor V.H. (Ed.) *Flowering Plants of the World* Croom Helm, London, 1985

Hill, J.E. and Smith, J.D. *Bats: A Natural History* British Museum (Natural History), London, 1984

Holm, J. *Squirrels* Whittet Books, London, 1987

Huxley, A. *Green Inheritance* Collins/Harvell, London, 1984

Isler, M.L. and P.R. *The Tanagers: Natural History, Distribution, and Identification* Smithsonian Institution Press/Oxford University Press, Washington, DC and Oxford, UK, 1987

IUCN 1988 *IUCN Red List of Threatened Animals* IUCN, Cambridge, UK, 1988

Kimber, G. and Feldman, M. *Wild Wheat: An Introduction* The Weizmann Institute of Science, Rehovot, Israel, 1987

King, P. *Protect Our Planet* Quiller Press, London, 1986

Laidler, K. *Squirrels in Britain* David & Charles, North Pomfret, USA and Newton Abbot, UK, 1980

Leakey, R.E. *The Making of Mankind* Michael Joseph, London, 1981

Lever, C. *Naturalized Birds of the World* Longman Scientific & Technical, Harlow, UK, 1987; *Naturalized Mammals of the World* Longman, London and New York, 1985

Little, E.L. *The Audubon Society Field Guide to North American Trees* Alfred A. Knopf, New York, 1980

Margulis, L. and Schwartz, K.V. *Five Kingdoms: An Illustrated Guide to the Phyla of Life on Earth (2nd Ed.)*, W.H. Freeman, New York, 1988

Moore, D.M. (Ed.) *Green Planet: The Story of Plant Life on Earth* Cambridge University Press, London and New York, 1982

Mourier, H. and Winding, O. *Collins Guide to Wild Life in House and Home (Trans. Vevers, G.)* Collins, London, 1977

Niering, W.A. *The Audubon Society Field Guide to North American Wildflowers (Eastern Region)* Alfred A. Knopf, New York, 1979

Perrins, C. *Collins New Generation Guide to the Birds of Britain and Europe* Collins, London, 1987

Secrets of the Seashore: The Living Countryside Reader's Digest, London and New York, 1984

Spellenberg, R. *The Audubon Society Field Guide to North American Wildflowers (Western Region)* Alfred A. Knopf, New York, 1979

Stanley, M. and Andrykovitch, G. *Living: An Introduction to Biology* Addison-Wesley, Reading, Massachusetts, 1982

Stott, P. *Historical Plant Geography* Allen & Unwin, London, 1981

Streeter, D. *The Wild Flowers of the British Isles* Macmillan, London, 1983

Thomas, B. *The Evolution of Plants and Flowers* Peter Lowe, London, 1981

Thurman, H.V. *Introductory Oceanography (5th Ed.)* Merrill, Columbus, Ohio, 1988

Udvardy, M.D.F. *The Audubon Society Field Guide to North American Birds (Western Region)* Alfred A. Knopf, New York, 1977

Webb, J.E., Wallwork, J.A. and Elgood, J.H. *Guide to Living Mammals (2nd Ed.)*, Macmillan, London, 1979

Whitaker, J.O. Jr. *The Audubon Society Field Guide to North American Mammals* Alfred A. Knopf, New York, 1980

Whitfield, P. (Ed.) *Illustrated Animal Encyclopedia* Longman Group, Harlow; Macmillan Publishing, New York, 1984

Zohary, D. *Domestication of Plants in the Old World: The Origin and Spread of Cultivated Plants in West Asia, Europe and the Nile Valley* Clarendon, Oxford, UK, 1988

Acknowledgments

ARTWORK CREDITS

Land-centered base map by Oxford Cartographers
Ocean-centered base map by Eugene Fleury
Line work and map overlays:
pp. 18, 19, 20, 21, 22, 23, 32, 33, 40, 41, 43, 44, 45, 56, 57, 79, 86, 90, 96, 97, 98, 100, 116, 117, 118, 119, 122, 137, 145, 146, 150, 153, 156, 157, 159, 161, 163, 176, 177, 188–199 by Technical Art Services
pp. 29, 43 (globe), 103, 106, 114, 125, 152, 157 by ESR Ltd
pp. 25, 30, 34, 35, 37, 46, 54, 55, 126, 138, 139, 156, 157, 166, 167, 180–187 by Ed Stuart

Other artwork

t = top; *b* = bottom; *l* = left; *r* = right 12–13 *t* to *b*: Ed Stuart, Michael Woods, Ed Stuart

14–15	Line & Line
16–17	Dave Ashby
18–19	Steve Kirk
20	Shirley Felts (artwork based on reconstructions published in *The Evolution of Plants and Flowers* by Dr Barry Thomas, Peter Lowe, 1981)
22	*t* to *b*: Dick Twinney, Graham Allen
23	Graham Allen
24	(*clockwise*): Graham Allen, Dick Twinney, Graham Allen, Graham Allen, Keith Brewer, Keith Brewer
25	Graham Allen
26–27	Dave Ashby
32	gorilla: Graham Allen; others: Tony Graham
33	Tony Graham
35	*t* to *b*: Graham Allen, Steve Hollen, Graham Allen, Andrew Robinson, Graham Allen, Graham Allen, Graham Allen, Malcolm Ellis
37	Ed Stuart
38	Karen Daws
39	Tony Graham
40	Tony Graham
41	bird: Michael Woods; skink: Alan Male
43	Michael Woods
44	Tony Graham
46	Ed Stuart
47	Ed Stuart
52–53	*t*: Shirley Felts; *b*: Karen Daws
56–57	Liz Pepperel
80–81	Paul Richardson
83	Ed Stuart
84–85	Liz Pepperel
87	Dave Ashby
88–89	Liz Pepperel
90	Tony Graham
96–97	Tony Graham
98	Graham Allen
100	Michael Woods
102	Paul Richardson
103	Graham Allen
104–105	Michael Woods
108	Graham Allen
109	*t* to *b*, *l* to *r*: Steve Kirk, Graham Allen, Graham Allen, Colin Newman, Steve Kirk
115	*t*: Paul Richardson; *b*: bird: Michael Woods; map: Eugene Fleury
116	Dick Twinney
126–127	Paul Richardson
128–129	Ed Stuart
130–131	Tony Graham
137	Andrew Wheatcroft
141	Paul Richardson
142–143	Paul Richardson
146	Karen Daws
148	Michael Woods
149	Paul Richardson
150–151	Dave Ashby
152	Dick Twinney
154–155	Paul Richardson
159	Paul Richardson
164–165	Shirley Felts
166	Ed Stuart

PHOTOGRAPH CREDITS

t = top; *c* = center; *b* = bottom

1	Harald Sund/The Image Bank
2–3	E. Svensen/Zefa Picture Library
4–5	Ulli Seer/The Image Bank
6–7	John Garrett/Tony Stone Associates
8–9	Margarette Mead/The Image Bank
15	David Paterson
17	R. Villarosa/Overseas/Oxford Scientific Films
21	Dr John Feltwell/Wildlife Matters
28–29	Shang Mingqi/Xinhua News Agency
31	François Gohier/Ardea
36	Soames Summerhays/Science Photo Library
40–41	David Ward/Landscape Only
42	Tony Rodd/Weldon Trannies
45	Frans Lanting/Bruce Coleman
47	Kevin Schafer/Peter Arnold
48–49	Steve Terrill
58–59	Liz & Tony Bomford/Ardea
60–61	J. Grant/Natural Science Photos
62–63	Günter Ziesler
64–65	E. Rekos/Zefa Picture Library
66–67	Hans Christian Heap/Planet Earth Pictures
68–69	David Woodfall/N.H.P.A.
70–71	Stan Osolinski/Oxford Scientific Films
72–73	Peter M. Miller/The Image Bank
74–75	Don Landwehrle/The Image Bank
76–77	Stephen Krasemann/N.H.P.A.
78	Spencer Swanger/Tom Stack & Associates
81	Milton Rand/Tom Stack & Associates
84	Nicholas Penn/Planet Earth Pictures
86–87	Tony Stone Associates
89	NOAA/NESDIS/NCDC/SDSD
91	Ron & Valerie Taylor/Ardea
92–93	Hans Wolf/The Image Bank
96	Wayne Lankinen/Bruce Coleman
99	Stephen Dalton/N.H.P.A.
105	Horus/Zefa Picture Library
106*t*	Favardin C./Jacana
106*c*	Eric Crichton/Bruce Coleman
106*b*	H. Xavier/Jacana
107	Dr C. Grey-Wilson
110–11	Pete Turner/The Image Bank
114–15	David Jesse McChesney/Planet Earth Pictures
117	Y. Arthus-Bertrand/Peter Arnold
118–19	Superstock International
121	Julia Sims/Peter Arnold
123	K.G. Preston-Mafham/Premaphotos Wildlife
124–25	Gianni Tortoli/Photo Researchers
127	Christian Zuber/Bruce Coleman
128–29	Smithsonian Institution
131	Denis-Huot/Agence Hoa-Qui
132–33	S. Wilkes/Zefa Picture Library
140	Tony Stone Associates
143	Roland & Sabrina Michaud/The John Hillelson Agency
144–45	Terrence Moore
146–47	Liz & Tony Bomford/Survival Anglia
148	Jen & Des Bartlett/Bruce Coleman
149	Biophoto Associates/Science Photo Library
151	A.P. Paterson/Ardea
153	Heidi Ecker/Horizon
154–55	Tom Stack & Associates
158–59	Bruce Davidson/Survival Anglia
160	Jeremy Hartley/Panos Pictures
161	Zefa Picture Library
162–63	Zefa Picture Library
168–69	R. Künkel/The Image Bank
170–71	P.B. Kaplan/Photo Researchers
172–73	James M. McCann/Photo Researchers
174–75	Daily Telegraph Colour Library
178–79	Peter Johnson/N.H.P.A.

Contributors:

David Attenborough
10–11, 50–51, 94–95, 112–13, 134–35

Barry Cox
14–29, 36–39, 42–43, 46–47, 108–9, 126–31, 136–37, 142–43, 152–53, 168–69, 176–77

Peter Moore
30–35, 40–41, 44–45, 52–79, 82–85, 96–97, 106–7, 122–23, 138–41, 144–47, 156–57, 164–67, 170–75, 180–83

Philip Whitfield
12–13, 80–81, 86–91, 98–105, 114–21, 124–25, 148–51, 154–55, 158–63, 184–87

The publishers would like to thank the following people for their invaluable help in the making of this book: Blackwell Scientific Publications Limited for their kind permission to base maps and diagrams on pp. 18–19, 20, 39, 47, 57, 103, 152 on material from Cox, *C.B.,* and Moore, P.D. (1985) *Biogeography*; David Carter of the Entomology Department, British Museum (Natural History) London, for supplying reference material; Judith Beadle for preparing the index.